Vassar Coly
Aug '92

The
JOY
Of
SCIENCE

EXCELLENCE
AND
ITS
REWARDS

The
JOY
Of
SCIENCE

EXCELLENCE
AND
ITS
REWARDS

CARL J. SINDERMANN

PLENUM PRESS • NEW YORK AND LONDON

Library of Congress Cataloging in Publication Data

Sindermann, Carl J.
   The joy of science.

   Includes bibliographical references and index.
   1. Scientists.   2. Research—Methodology.   I. Title.
Q147.S56 1985   502′.3   85-6604
ISBN 0-306-42035-X

© 1985 Carl J. Sindermann

Plenum Press is a division of Plenum Publishing Corporation
233 Spring Street, New York, N.Y. 10013

Printed in the United States of America

PREFACE

The vague outline of a book based on case histories of successful scientists emerged slowly in discussions with colleagues during a decade of exhausting research associated with the creation of an earlier book titled *Winning the Games Scientists Play*. The perception grew to overwhelming proportions that success in science is closely linked to pleasure in the practice of science, but that this simple observation had rarely been emphasized within or outside scientific circles.

Since these insights could not be encompassed satisfactorily in the earlier book, the logical solution was to develop a second book emphasizing the broader horizons of successful careers of scientists. A few sample chapters were drafted and floated past carefully selected colleagues. The response could not be described as wildly enthusiastic, but good sense dictated that the project be continued. Extensive field research was necessary to test the basic premises and the initial perceptions. The best data and the most brilliant insights were gained during late evening cocktail parties at professional meetings. Other good insights were acquired during "twilight sleep" while particularly boring scientific papers were being presented in dimly lit session rooms at those same meetings.

However, much of the factual base for the book was accumulated through detailed conversations resulting in oral case histories gathered over a period of several years. The present document thus has a reasonable quantitative foundation, although its factual base could hardly be described as "statistically robust."

As the following pages will attest, this book is clearly, obviously, and emphatically not an attempt at autobiography; rather it is a tale told by an innocent, sometimes gullible observer of scientists in their preferred habitats. It is an account of some of the reasons for success in science, based on extensive scrutiny of those who seem to be contributing to knowledge and enjoying their profession.

Books of this kind are of course based on perceptions—which can vary greatly from one observer to another. This is a sagittal section of the scientific world as seen by one inhabitant, with its successes and its joys in neat arrays. Similarities to perceived worlds of other analysts may be only coincidental.

I want to thank all the brave but securely anonymous scientists who contributed to the conceptual and factual base of this book—by example or by prolonged discussions, usually long after their normal bedtimes.

CARL J. SINDERMANN

Miami, Florida

CONTENTS

PROLOGUE

In the mid-1970s two remarkable books about scientists were published—*The Visible Scientists* by Rae Goodell, and *Scientific Elite* by Harriet Zuckerman. Goodell's book focused on eight contemporary "public scientists," while Zuckerman's was concerned with Nobel laureates. *The Joy of Science* examines another stratum of scientists—the very good, the excellent, and the near-great scientists who make up the vastly larger professional core of scientific disciplines. Unlike the highly selected samples in the two books mentioned above, the excellent scientists scrutinized in this one are so abundant that description of even the most obvious categories is a major undertaking. Accepting the maxim that "Fools rush in . . . ," this book attempts to identify some characteristics of successful scientists and to describe the pleasures and joys (and even some of the sorrows and agonies) of existing within the scientific community.

This is an account of those many excellent scientists who may never get invited to late evening television talk shows or be nominated for a Nobel prize—but who are the men and women making significant contributions to science. Much of the book is frankly and overtly "elitist" in that it focuses on those who "make it" and why they do so. Some attention is given to the "also-rans," but only for contrast. The goal here

is to identify and describe the "excellent scientist"—in universities, government, and industry—and explore some of his or
her job satisfactions. Careers are examined for common and/
or unique characteristics which may help to circumscribe the
nature of excellence and the reasons for success, within a scientific discipline and in interpersonal relationships within the
sciences. Valid insights into these matters can be elusive, conclusions may be disputed, and analyses might be considered
superficial or naive—but the *search* seems to be a legitimate
objective.

An underlying thesis of this book is that some commonalities do exist in successful scientific careers; that these can be
elucidated after appropriate analyses; and that a consistent
experience during most professional life spans is "joy"—a
sense of internal well-being and satisfaction. A hope is that
good scientists—wherever they are—will find small reflections
of themselves and/or their experiences in some of these pages.

The research on which this book is based started out as a
simplistic search for characteristics of excellence in science,
which should lead to success in science, which should in turn
lead to joy in the practice of science. This analytic sequence
proved to be a trap, recognized too late for orderly retreat. The
terms "excellence," "success," and "joy" are to many scientists
overlapping, controversial, and undefinable. In discussions
with colleagues, no one questioned the importance of the
vague concepts lurking behind the terms, but no one agreed
with my attempts to encircle them.

Such obvious conceptual disorder provided the impetus
for an excursion through abstractions like "good science,"
"succeeding in science," "scientific credibility," "scientific
authority," and "joys of science," as well as through some of
the less attractive aspects of science—fraud, controversies,
fade-out, and senescence. This trip should be like a journey
with a too-brief guidebook through some of the great cathedral

cities of Europe, following an itinerary which includes, in addition to the tourist attractions, an obligatory afternoon spent in the low-rent districts of each one.

Periodically in the following chapters attempts will be made to define some of the terms used; this prologue seems like a logical place for a first assault on them. An entry-level definition of "success" is "that professional state which results from significant contributions to human understanding of natural phenomena, recognized by peers, and internalized by the contributor." Those who "make it"—defined as "those individuals who emerge above the level of the average, through instinctive and learned professional activities"—do so because they practice "good science," encapsulated as "production of extensive data, rational insightful application of those data to hypotheses about natural events, and effective presentation of the resulting information and analyses to colleagues."

A foundation of excellence in the *practice* of science constitutes the core of success, around which additional interpersonal embellishments may be added, but which can never replace the core. Elements of the core are correct choices of research problems to be addressed, logical experimental design, sustained productivity, significant insights based on interpretation of research results, contributions to the conceptual framework of a discipline, knowledge of the developing literature in a chosen field, effective communication of findings in professional forums, and an occasional unostentatious spark of brilliance in one or more of the preceding.

The interpersonal embellishments of the core include (but are not limited to) good rapport with colleagues involved in significant research in a specialty area, personal recognition as a key figure in that specialty, active involvement in one or more professional societies, frequent participation in workshops and symposia, and contacts with colleagues in other countries.

Although there may be an element of wishful thinking on my part, this book seems to have some elusive cohesiveness, supplied by continued attention to the interlocking triumvirate of excellence/success/joy. Four principal sections provide some substance and even a few facts. The first section, "Getting There," focuses on some important components of the upward climb of those who have achieved success in their fields—how they formed a productive research team, how they structured investigations, how they presented findings, and how they managed career transitions. The second section, "Scientific Elite," is really the heart of the book. It scrutinizes the various roles of scientists—research, teaching, administration, and others. It considers some other "hows" of success; it describes authority figures in science; and it even singles out female scientists for special attention. The third section, "Descent from Career Peaks," provides relief from too much success and joy by exposing the unattractive underbelly of science—the also-rans and the failures, and where they went wrong. The final section, "Encounters with the External Environment," moves scientists out of their hiding places in laboratories and classrooms into public forums.

Traveling along these various routes we can expect to encounter scientists in transition, mentors and other authority figures, scientists in search of immortality, and even some guaranteed losers. The title promises "joy," and I hope joy will be delivered, along with an occasional driblet of realism.

It should be emphasized from the beginning that science is conducted against a stage setting of the personal lives of scientists, about which this book, regretfully, will have little to say—not because personal lives are unimportant, but simply because they introduce an impossible level of variability—far beyond the author's marginal competence in psychology, sociology, and statistics. Available data, as most of us know very well, can be massaged and stretched only so far; the objective here is to examine success and joy in the practice of science, and not to become enmeshed in too many nonscientific components of a total equation for living. An early reviewer of drafts of this book tried to point out to me that any reasonably complete analysis of the careers of successful scientists should include personal backgrounds and life crises, insofar as they affect professional activities, and failure to include more of the human side of scientific careers denies the reader many insights about the "highs" and "lows" of human existence. I rejected the comment, claiming to have my hands full just trying to examine the professional side of things—and the satisfactions to be found on that side of the street. I remain convinced that within the career microcosm called "science" there are pleasures which transcend events in the outside world—and which can provide positive reinforcement for the normal joys of living.

PART ONE

GETTING THERE

Career journeys in science require at least a minimum of essential baggage—reasonable intelligence, an interest in problem-solving, and a curiosity about natural phenomena. Accessories could include an advanced degree in a specific discipline, a mentor who expects star performances, and a spouse or good friend who keeps repeating with some sincerity, "You're great," even when the evidence is not yet convincing. Additional but less critical items are supportive colleagues, a job which has scope for scientific growth, and friends in the institutional hierarchy.

Even a full complement of these good things will not put a beginner in the winner's circle; work has to be done in the tile-lined saunas of scientific laboratories—honest, necessary, occasionally exhilarating, but usually monotonous and repetitious work, commonly called "data acquisition." Joys in such work lurk in the rare insights, the completion of an exhausting test series, the statistical tests which indicate validity and definite correlations, the good midnight discussions with sharp colleagues, or the infrequent lost weekends between experiments when it is proper and even necessary to get as far from the lab as possible without leaving the planet.

Beyond these routine pleasures are the real prizes—the synthesis that eluded earlier investigators, the simple but elegant and definitive demonstration of causality, the beautiful smile from a normally dour technician, or the firm congratulatory handshake from an autocratic, demanding mentor. Only then does the newcomer begin to think that good science is almost within reach, and that maybe there is some fun in doing good consistent research.

A long-time associate, and an excellent scientist in many ways, Dr. J. C. Sunderland, currently with the Wade Institute for Deep Ocean Research in Greenbackville, Virginia, put this vague feeling of incipient worth into words recently. He described it as a rare precious moment when a new scientist first recognizes that he or she is touching—however briefly—an area previously unexplored.

The first major section of the book tries to trace those early somewhat uncertain steps of the scientific novice whom we have later known to grow and to achieve recognition and respect from colleagues. A description of good science, even at an entry level, is difficult but necessary, because it is from the ranks of the naive, pig-headed, brash, but very bright upstarts that the "authority figures" and prize winners of the next decade have emerged and will continue to do so. This section, then, attempts an *a posteriori* analysis—looking backward at the beginnings of what later proved to be successful careers. A note of caution might be inserted here, though, that this type of analysis is not without a major flaw—the difficulty of predicting whether certain elements common to careers of successful scientists will be forecast reliably in novices. Despite this flaw, the elements have some diagnostic value, and are the best indicators that we have available.

CHAPTER 1

MOVING UP

The "Method" of the Scientific Method; The Origin of Research Ideas; Experimentation and Observation; Conclusions and Syntheses; Communicating Research Results

INTRODUCTION

Dr. Arthur G. Hoechst,* a flamboyant aggressive near-genius, backed by a large research team, is a certified "authority" in the expanding field of genetic modification; Professor Walter W. Woodside, a superb organic chemist, diplomatic and politically astute, participates actively in several panels of the National Academy of Sciences; Dr. Lemuel R. Groveland, introspective, introverted, and laboratory-oriented, has just received a regional award for contributions to understanding the nutrition of wood-eating cockroaches; and Dr. Alan W. Castell, high-energy physicist, has ongoing projects and col-

---

*The names of most of the scientists in this book are fictitious, but the anecdotes and vignettes are drawn largely from the real world and not a world of fantasy—with some modest deviations from fact to prevent too-easy associations. In the case of the use of an actual name, it will be designated as such.

laborators at research centers in several countries. These four people are superficially disparate in every way but one—they seem to be succeeding as scientists.

What, if any, are the common denominators in their early professional development which might have given clues about their subsequent success—and can these same identifying traits be recognized in careers of other successful colleagues? Answers to such difficult questions are buried in the untidy masses of data on which this book is based. Some insights may be derived, however, from scrutiny of how emerging and emergent scientists get ideas, how they design and conduct investigations, and how they analyze and present results of their work. These are the foci of this first probe into the universe of the successful scientist.

THE "METHOD" OF THE SCIENTIFIC METHOD

Beginning in some chart-bedecked secondary school science classroom, a simple superficially logical concept, "the scientific method," is usually proposed to unsuspicious junior minds as an explanation of how scientists conduct their business. Classical steps in the "method" form a readily memorized list (with some variations, depending on the teacher):

- Formulation of a hypothesis
- Accumulation of relevant data through observation and experimentation
- Possible modification of the hypothesis based on interpretation of data
- Further observation and experimentation to verify the revised hypothesis
- Synthesis of all available data; and finally the capstone—
- Statement of a concept

What a beautiful blueprint for action! What an orderly way to search for truth! What a fraud! Most successful scientists would agree with Broad and Wade in their book *Betrayers of the Truth*[1] and even with Feyerabend's treatise *Against Method*[2] that there is no single specific "scientific method"; that it is a philosopher's invention imposed on the everyday world of scientific research, with little foundation in reality. Those same scientists would probably agree, however, that *there is a state of mind, an approach to problem-solving, that is common to scientific observation*—and that *this* is the essence of the methodology of science. Common elements include objectivity (insofar as the subjective mind will permit it), insurance of adequate controls, dispassionate analyses based on adequate statistical treatment, and assurance of validity of samples. Conclusions must not extend beyond the data on which they are based (the foundation of a delightful modern EPA concept of "legal viability").

Reality, for most professionals, is far sloppier than the neat textbook "scientific method," and follows no single pathway. The evolution of and the progressive refinement in methods and concepts may be more acceptable explanations of how science is done. The process includes

- *Evolution of ideas and insights*, from the first faint hint of something on the horizon, through repeated blind alleys and diversionary channels, to a final testable statement
- *Evolution of experimental design*, from crude "let's just see what happens if we do this" to an elaborate, successive-step, equipment-intensive series of progressively complex experiments, to the final exquisite, definitive, but superficially simple demonstration
- *Evolution of data analyses*, from the first rough correlations to testing of elaborate computer-based models

- *Evolution of syntheses*, based on data analyses, incorporating relevant conclusions and insights published by other investigators

The scientific method, therefore, can be best visualized as a succession of stages in the progression of thoughts, complete with major or minor changes in direction and with nodal points where critical insights have occurred. It is much like a treasure hunt in which the original designer of the course has disappeared (or is at least unavailable for direct consultation).

If these limitations on the reality of a single "scientific method" are accepted, it is still possible to track the emergent scientist through the critical stages already mentioned—by examining (1) the origin of research ideas, (2) experimentation and observation, (3) conclusions and syntheses, and (4) communication of findings.

THE ORIGIN OF RESEARCH IDEAS

Some research ideas are original—really creative—but much of scientific progress does not depend on unique concepts; rather it depends on filling in blanks, caulking gaps in data, and occupying tiny niches in a partially completed mosaic. My collaborator and faculty club bar companion, Professor J. C. Sunderland, whom I mentioned earlier, tried once, in the early days of our association, to assemble for me a partial list of sources of ideas for novices. Although his roots were in a simpler long-dead era of science, and although he was a dedicated pontificator, some of his expositions are relevant today:

- For graduate students, the primary source of ideas is the major professor, who may have a sack of them, well picked over by previous graduate students and a little shopworn, but possibly with a few remnants of gem quality, if properly polished. Some of the professor's ideas may have developed from his or her ongoing research, and some small tidbits of these may be doled out and considered.

- Seminars or session papers at professional meetings have proved to be excellent places to begin extending the conclusions and ideas of others to new research. Often just being in an environment where science is discussed will set thought processes in motion, some of which may be totally unrelated to the topic of the presentation.
- Reading newspaper accounts or popularized magazine articles on new findings in related fields may set off trains of thought which lead to research ideas.
- Doing a literature search, preparing for a lecture or seminar, may disclose a conspicuous gap in the available data, and may motivate a more serious search, and eventually a research proposal.
- A blank pad of paper on an otherwise empty desk has been for many an excellent way to encourage thought processes about future research.
- An undergraduate asking a stupid question, for which no answer is available—stupid or otherwise—has been known to set scientific thought processes in motion.
- At a public lecture or panel discussion, an industry member, a housewife, or a representative of an activist group may ask an embarrassing question, for which there is no reasonable response except "I don't know, but I'll look into it."
- A late evening barside discussion with colleagues may suggest holes or discrepancies in existing data—which should be filled or corrected quickly.
- While gazing vacantly from an office window at a soggy November campus, or during the dismal early morning

hours of a sleepless night, a true "eureka episode" of sudden insight may occur (but only very rarely).

This is a pitifully inadequate list of some of the ways that professionals have acquired ideas that enabled them to go about "doing good science." Any practitioner could develop a comparable list with little duplication of items. Scientists attend meetings, and hear about techniques which could be applied to their own work; they scan scientific journals and find papers which summarize material relevant to their research; or they thumb idly through journals in a related field and find throwaway ideas in discussion sections relevant to their studies. Professionals, as distinct from amateurs, must have the background, the judgment, and the intuition to evaluate ideas and insights—accepting or rejecting them for further action. Background and judgment enter also in the formulation of hypotheses and the development of experimental designs with statistically defensible foundations. Data collected to provide support for or to refute a hypothesis must result from well-planned experiments or observations—no seat-of-the-pants estimates, no hip-boots or butterfly-net trial-and-error sampling will suffice for the professional.

EXPERIMENTATION AND OBSERVATION

Interwoven with ideas and insights—and difficult to distinguish sharply from them—is the conduct of investigations through experimentation and observation. Ideas and insights are of course the principal foundations of research, but they must be accompanied by action—by data collection and analysis. The ideas and the action follow parallel paths with frequent crosswalks and mutual reinforcement.

Dr. Arthur G. Hoechst, mentioned at the beginning of this chapter, once pointed out to me that some of the satisfactions of the action phase of doing good science include experimental series that flow smoothly toward well-defined end points, field observations that fit nicely with laboratory studies, and methods adapted from previous studies which prove workable in new research. He was quick to emphasize, however, that these satisfactions are extremely rare, and that the norm tends to be more often in assorted frustrations—of experiments which lead to totally ambiguous results, of field observations which are anomalous and confusing, or of methods adapted from previous studies which prove absolutely unworkable in new applications.

Many respected professionals have over the years developed a series of unwritten guidelines or admonitions which they heed almost unconsciously in their own personal research. The list could be extensive, but it would probably include some or all of the following:

- Have a research plan or strategy—long-term as well as short-term.
- Read the literature—including foreign journals or abstracts—to know where the frontiers are.
- Select problems of substance that seem amenable to solution within tolerable time frames; consider all possible hypotheses, but begin by testing the most reasonable ones.
- Do pilot or exploratory work on a modest scale to expose shortcomings in techniques and other problem areas, before quantitative studies begin.
- Collect more data than will be needed, but determine in advance what a statistically valid sample is.
- Avoid "unconscious advocacy" in personal research, in which experiments and data favoring a particular conclusion may receive undue weight in evaluating a set of results.

- Remember that replication is an essential feature of experimentation; nonreproducibility of results can provide clues about intrusion of extraneous or noncontrolled factors.
- Expect the definitive experiment to emerge slowly; it will not usually jump from behind the first door.
- Make it an absolute rule that if several laboratories are participating in analyses, intercalibration and standardization of techniques must be done.
- Be prepared, when the weight of negative evidence dictates, to discard or modify a hypothesis—even though it may have grown to be a pet.

Most scientists, if pushed, would probably admit to the existence, at least in their minds, of such a list of guidelines, if only as self-help measures in experimentation and observation—which are the critical links between ideas and the demonstration of reality.

CONCLUSIONS AND SYNTHESES

It is hard to generalize about which part of the process of "doing good science" is most pleasurable to participants. For some, the data-gathering phase, with the hoped-for emergence of insights, is the most exciting; for others, that phase is a bore, and only a means of getting to the important work of analyzing the complete data set, drawing conclusions, developing syntheses with other available data, and interpreting the findings.

The pleasures of this analytic integrative phase of research are in seeing trends, in demonstrating valid correlations, and in perceiving solutions to problems. It is here, in this analytic phase, that professional excellence will be most apparent. The

average scientist will see facts as facts, and may make reasonable interpretations of collated data; the excellent scientist will
see the same data as parts of a mosaic of an intricate design,
which he or she will perceive early and with great clarity. The
average scientist will use standard methods of analysis; the
excellent scientist will often do something genuinely innovative. The average scientist will feel relief that a unit of work
has been finished; the excellent scientist will be uneasy and
excited about the next step in the research.

Reading all this, one would be reasonable in asking, "How
much of science is truly creative?" The answer is, "Less than
intuition would lead one to believe, since so much of science
builds on research that has been published or discussed or
reported previously." Of course, there are moments of insight,
tiny flickers of new syntheses, new approaches to persistent
problems, and new techniques, which lead to additional perspectives on those problems and add to the structure of knowledge. Also, there are many degrees of "creativity"—from the
rare exploration of a genuinely original idea to the pedestrian
development of a data set which augments others already in
existence, and which may contribute almost imperceptibly to
validation or refutation of an existing hypothesis.

The genetic manipulation of bacterial syntheses, the
recent progress in understanding the nature of subatomic par-
ticles, the laboratory synthesis of complex molecules—all are
heralded as major advances in research, and they are. Yet such
advances do not descend on humans like lightning bolts; they
are built on thousands of hours of thought and labor by countless research people, many of them now retired or dead, and
most of them not participants in any of the honors bestowed.
Tracing the history of contributions leading to any single major
advance becomes almost an exercise in futility, if one were to
include each step and each contributor, direct or indirect, to the
emergence of a concept. How, for example, do we recognize

and include the developers of equipment and techniques used to separate serum fractions which can then be used in purification of antibodies which can be used to detect a bacterial strain useful for genetic engineering research? The actuality seems to be that scientific progress does not occur in a vacuum—it is built on analyses of newly acquired information (and conclusions drawn therefrom), but also on all the information available to the investigator through publications and oral presentations of research findings by others.

COMMUNICATING RESEARCH RESULTS

The good, well-designed, statistically sound research has been done; the data have been collated, analyzed, and interpreted; the graduate assistants have been paid; and the grant summary has been submitted. End? Not on your life! Presentation of findings in major journals is an additional, absolute, no-exceptions requirement. This step, inexplicably, is the one which separates the men from the boys, the women from the girls, the professionals from the dilettantes, the winners from the also-rans.

So much has been said about good written scientific presentation of research findings that further discussion here would seem redundant. Not so! Perusal of recent volumes of scientific journals suggests that we take one small step for science and one giant leap for journal editors, in reaffirming a few basic rules of the road in scientific writing.

Professor Lemuel R. Groveland, mentioned at the beginning of this chapter, served as editor of a zoology journal, and spoke knowledgeably (if not charitably) about the writing of novitiates and cardinals in the scientific hierarchy. His message was "We should do better!" How we do better, he said, short of calling on ghostwriters, is to seek out and follow the rare examples of good scientific writing, wherever they can be isolated and identified. From such examples, he had extracted over a period of years some excellent rules of conduct in written communication (summarized here with his permission):

- Every step in the conduct and reporting of scientific research—design of experiments, testing of results, and formulation of conclusions—must be subjected to statistical examination as a stamp of validity. Unfortunately, statistical treatment of data is itself subject to perceptions of individual scientists. Each subdiscipline seems to have its own specialized quantitative measures. Some standard methods, such as chi-square, persist, but more elaborate analyses quickly become badly fragmented among numerous mathematically oriented cults. Statistical methods used in any paper may therefore be either applauded or condemned by reviewers who are practicing members of one or another sect.

- Critical components of scientific papers tend to cluster in the hindmost segments—in the consideration of results, conclusions, and interpretations. The results of an investigation should not be intermixed with discussion of those results or of the results obtained by others; this is a major failure in many papers, and may be a result of deliberate obfuscation in some few cases. Results are usually stated in a section of the paper so labeled, but too often the final step—conclusions to be drawn from the results—is left floating at the end of a section of the paper labeled "Discussion," or is incorporated unsatisfactorily in capsule form in the "Summary." Much more precise, and certainly more understandable, is the creation of a subterminal section called "Conclusions," and there to state clearly the author's thinking about the meaning of the work. This imposes a

burden on the author, who is constrained to stay within
the confines of the data.

- A beautiful but physically abused ingredient of a scientific paper is the so-called Discussion. The purpose of such a section is to place the results of the reported research within the larger structure of understanding of concepts or principles. It can be a vehicle for unobtrusive attempts to indicate contributions of the present study to established or developing areas of understanding. It can also be a combat zone where deficiencies or incorrect conclusions of other published work are discussed.

Authorities in some subdisciplines of science sometimes achieve notoriety among colleagues through use of "Discussion" sections of their papers as clubs—to beat on upstarts who dare to publish in their domain, or who tamper with accepted concepts. One of my all-time favorite practitioners of this genre of writing is Dr. Laurence H. Honeyspot, a prolific researcher in the geochemistry of deep ocean sediments. This is his turf, and he protects it vigorously. Any author who strays into that territory and publishes findings has to be very good or else he or she will be pounced on and shredded expertly in the Discussion section of Honeyspot's next paper. Those of us familiar with his work and his idiosyncracy await his next paper principally to see who among his juniors will be next to receive attention. Editors of journals in which he publishes condone some of this, since he is so obviously superb as a scientist. Spice was added to the game recently when one very sharp junior colleague began striking back, using Honeyspot's tactics of nitpicking, then shredding. This is hard to do with Honeyspot's excellent papers, but no one is perfect. Honeyspot is, however, perceptive as well as good; he completely disarmed and silenced the junior by offering to co-author a series of papers with him—an offer which the junior accepted, and which materially advanced his career. There is still the paradox, though, of an excellent mature scientist diminishing his reputation by these turf-protecting mechanisms in scientific papers.

Some additional conclusions about scientific writing proposed by Professor Groveland are these:

- Excellent scientists are careful not to allow a "Discussion" section to become a "free zone" for trade in weakly supported speculations or to serve as a vehicle for a review of the entire field of specialization (unless it is so titled). It is legitimate to explore topics of direct relevance to the subject of the paper—no more. Most editors become increasingly restless as the "Discussion" drags on; some set arbitrary limits such as "The Discussion section will not exceed the Results section in length," or "The pages of the Discussion section will not exceed one-fourth the length of the entire text." Some journal editors are even driven to practices used by newspaper editors—they arbitrarily chop off a discussion when a stated maximum number of pages or words is reached. This method of pruning is usually reserved for stubborn authors who are reluctant to give up a single paragraph of their creation, yet they may have worthwhile results to report.
- Scientific papers should be objective reports of facts and conclusions derived from experimentation and observation. Authors occasionally yield to temptations to make them more than that—especially in writing the Discussion sections of manuscripts. Hazardous practices include "interpolation" (inserting assumed data points in incomplete series) and its big brother "extrapolation"

(extending data points beyond those determined by experiment). Having succumbed to these practices, some authors also indulge in conjecture and speculation far beyond the legitimate confines of their data—becoming increasingly vulnerable to being very wrong. Good writers state reasonable conclusions based on statistically sound data; they may even discuss the relationships of their findings to those of others; but they do not become essayists or reviewers as part of the same exercise.

As a postscript to this discussion on communicating research results, *oral presentations* should be given some visibility. Doing good science and reporting results in peer-reviewed literature are the big objectives, but standing up before colleagues to present those results orally is also an art form to be cultivated with great vigor by an emerging professional. The beauty and grace of an exceptional oral presentation, based on good data, are not lost on colleagues used to standard poorly orchestrated performances. We have all encountered—those of us who have served our time in purgatory during interminable hours spent in scientific sessions—those rare excellent papers, crisply and enthusiastically presented, with good organization, meticulous diction, outstanding visual aids, and significant findings to report. We try to applaud in a more than perfunctory way; we make a mental note to congratulate the speaker if we encounter him or her at the cocktail party; and we have a little better feeling about being in the presence of professionals.

Of even greater significance from a professional viewpoint, however, are the positive effects of good oral presentations on careers. Whereas success in science does not depend on oratorical skills, fellow scientists impressed by a well-presented substantive paper may at some future time be in positions to influence decisions about the colleague doing the talking.

A recent New York Times Magazine *article (August 1984) hailed the return of political oratory as an art form to be cultivated and appreciated. Can "scientific oratory" be classified similarly? Most excellent scientists would probably agree that the* content *of a scientific paper is of overriding importance, but few would disagree that a touch of the orator's art in presentation is appropriate too, provided it does not distort or overemphasize. One of my all-time favorites among those scientists seriously concerned with good (even at times dramatic) oral presentations is Professor George Welty, an excellent biochemist and a master at public speaking. His presentations at professional meetings are classics. He presents just enough relevant background to orient the participants, then he plunges enthusiastically into describing results of exquisitely designed experiments. Few but superb are his visuals, and his summaries are precise and concise. On occasion he even manages to insert brief episodes of carefully chosen humor, or unobtrusive laudatory comments on the research of others. The entire performance seems uncontrived, but those of us who know him realize that we are witnessing the outcome of many hours of painstaking preparation for every paper that he presents.*

Why, then, if oral presentations are an important part of growing up in science, are they usually treated so shabbily by most practitioners? The answer in part is that *they are rarely treated that way by excellent scientists,* even those like Dr. Groveland (mentioned earlier) who are by nature far from extroverts. They believe that effective oral communication is a component of excellence. The attitude they convey is that the best people never cease polishing skills. Hence, every meeting, symposium, or seminar is a learning experience—by observing brilliant performances, by squirming through awkward performances, or by presenting papers as often as possible.

It has become obvious to me, from long observation of excellent professionals, that once the art of good straightforward oral presentations has been mastered, acres of additional playing room become available to enhance interest and effectiveness—by going beyond the expected and the routine. The techniques require skill and practice; some of the ones that I have seen exploited successfully by excellent scientists at recent meetings are

- *The "midstream canoe change."* The presentation begins with a serious statement of a barely tenable or untenable generalization or assumption; then partway through the talk it is destroyed and replaced by one that the author really wanted to make.
- *The "bootstrap" Socratic approach.* A series of increasingly complex questions is asked, and then answered in sequence by the author. Each question arises from the previous answer.
- *The "cart/horse/cart" maneuver.* The presentation begins with the conclusions and generalizations, then supporting data are provided, then the generalizations are restated.
- *The pedagogue.* The presentation begins with a series of six true–false questions related to the material to be discussed, with a request that participants jot down T or F answers. The conclusion of the presentation is woven around providing the correct answers to the quiz.

Some of these techniques are not of course applicable to the standard 15-minute paper given at national society meetings, but wherever there is program flexibility, innovation is worth the slight risk of failure, or of confusing or enraging the audience.

Communicating research findings, then, can be an art form to be mastered by excellent scientists. We have all read exceptional scientific papers which contain results of good science presented logically, concisely, and interestingly; we have all listened to oral presentations so good that they almost bring tears to the eyes. Usually our colleagues capable of these masterpieces know that they are good, but they also know that being good is mostly a consequence of maximum investment and long preparation.

CONCLUSIONS

This attempt to describe some components of the working world of the journeyman professional seems to turn up a lot of effort and even more routine tasks. Where is all the joy of science that is advertised on the dust jacket of this book? The answer to this naive question must be that the early phases of a scientific career have to be characterized by hard grinding labor, the results of which may lead to the announced pleasures if the work is done diligently and brilliantly. Joys don't necessarily come early or easily, but they are out there waiting to be earned, then savored.

The early phases of career development have proven to be times for

- High-energy contributions to a project
- Much better than average insights
- Long, sometimes lonely hours in the lab or office, with at least part of it spent thinking
- Working and reworking an experimental design
- Agonizing over statistical treatment of data

- Incessant, almost compulsive, reading of journals, reprints, and professional books

Fortunately, even during these early days there are moments of what has to be described as joy, when a pattern of results begins to emerge, or when correlations are demonstrated statistically. My former mentor, Professor Addison J. Pead, an exceptional parasitologist and philosopher of science, tried to explain this to me long ago. It was, he said, a momentary high for him when some pattern began emerging from data, bringing with it a feeling of being on the edge of something really exceptional—which with one small additional effort, one more insight, one more mental struggle, would make a little piece of the universe forever easier to comprehend.

REFERENCES

1. William Broad and Nicholas Wade, *Betrayers of the Truth* (New York: Simon & Schuster, 1983).
2. Paul Feyerabend, *Against Method* (London: Verso, 1975).

CHAPTER 2

MANAGING A SUCCESSFUL RESEARCH GROUP

The Importance of Interpersonal Relationships to Successful Research; The High-Risk Area of Decisions about Authorship; Managing Peer–Colleague Relations

INTRODUCTION

The preceding chapter explored some basic operational approaches to science—the origin and refinement of ideas, the design and execution of experiments, the analysis and presentation of results—but a second principal ingredient of success must be explored early in any examination of scientists. This is the less structured area of interpersonal relationships in conducting science, particularly the delicate critical steps in building and managing a good research team.

It is almost axiomatic in the present climate of large research projects that much of science today is a team operation, consisting of a principal scientist and a supporting cast of postdoctoral assistants, graduate students, technicians, and secretaries. Sometimes these teams can be aggregated into

21

loose temporary cooperatives, to compete for the bigger multidisciplinary grants; and sometimes several principal scientists
will find compatibility and productivity as members of a single
research unit, sharing the support staff.

One fact emerges clearly: whatever the organizational
arrangement, *skill on the part of the principal scientist in the care
and feeding of all members of the group is critical to its success
and productivity.* This may seem too obvious and trite to
require emphasis here, yet the maxim that eludes many otherwise competent scientists is that *they must be very good at
managing a group.*

Superficially, it would seem enough to plan the research,
delegate responsibilities for its execution, and analyze the
results. Successful scientists recognize early, however, that this
is merely the framework for a complex and continuing relationship with people, who bring all their own priorities,
moods, and idiosyncracies to the laboratory each morning,
along with their competencies. Postdocs may depart for better
jobs on short notice, graduate students may be evicted from
their apartments, critical instruments may fail repeatedly, or
critical supplies may not arrive and may have to be borrowed
from colleagues. Coincident with all the daily internal crises,
grant review teams may descend, the administration may
screw up or question the budget, an undergraduate exam must
be graded, or a spouse may total the family car. How can
meaningful research be conducted in such a maelstrom? An
answer is elusive indeed.

---

*A conclusion based on years of casual or more than casual
contact with university research groups is that their management is so highly individualistic as to defy characterization. Among the few constants seems to be informality combined with rigid adherence to established procedures—an
odd but workable combination. Professor Harold G. Jacobs-*

son, Harvard Ph.D. and faculty member at a large eastern university, is a superb exemplar of this approach. A bouncy, gregarious, perceptive, productive scientist, he juggles lectures, travel, and management of several major research projects successfully, assisted by several postdoctoral assistants, two technicians, and a stream of graduate and undergraduate assistants. Consistent good humor, interest in assistants as people, and tremendous energy are combined with expectation of excellence from everybody on the team, and absolute insistence on careful experiments, fully replicated. Hal is one of a special breed who always has time, who listens to a problem, who sees a solution, and who enhances the joy of science for everyone he contacts.

The principal scientist is an analogue of the circus ring-master, preventing the onset of chaos from moment to moment, while at the same time presenting a facade of calm and efficiency. A combination of intrinsic abilities and learned techniques—appropriately applied—distinguishes the excellent professional from the mediocre one.

Scientists have known for a long time, but have been very careful to conceal the fact, that behind many successful careers are very competent and enduring laboratory technicians or research assistants (male or female). Most technicians, if asked, would assert flatly that scientists succeed because of their support group. If this represents a true state of affairs, then proper selection, adequate recognition, and close day-to-day interaction with technicians and assistants must be matters of utmost importance to successful scientific careers. Why then are scientists so often stupid or uncaring about such a critical part of the foundation of their success?

We all have our favorite stories about technicians who make scientists what they are today. One of my favorites is about a relationship which spanned 20 years in the lives of an out-standing chemist, Dr. M. M. Fox, and a superb technician, Miss Marlene Lindquist, who was recruited from a junior college program. She had marginal training for the complexities of chemical research, but was very intelligent, a reader of professional literature on her own time, and a model of efficiency and speed in laboratory procedures. Dr. Fox perceived very quickly just how fortunate he was to have acquired her, and pushed her into training courses and salary increases. Publications literally streamed from the pair, with Miss Lindquist as junior author on most and first author on a few. Somehow the pitfalls of a relationship that went beyond propriety were avoided (although the more evil-minded among us had our suspicions at times). Eventually the research group expanded slightly, with the addition of two junior male technicians, who found out immediately that she was clearly in charge of laboratory procedures, even though Dr. Fox was doing the planning and much of the writing when his international travel commitments permitted.

Those of us who have had the usual successes and failures in technician management marveled at and secretly envied this fantasylike relationship. Dr. Fox explained it matter-of-factly as unwaveringly professional and mutually supportive, with clear definition and acceptance of each role—scientist and technician—plus more than a little good luck on his part. I had hoped for a more enlightening response; he may not have told the complete story, but whatever it is, it works.

My former mentor, Dr. Addison J. Pead, felt so strongly about the importance of effective research groups and the amateur ways of most principal scientists in the management thereof that he designed and taught a seminar course for graduate students which he at first called "Group Dynamics in the Ivory Tower," but later changed to "Transactions Across the Laboratory Bench." The course was interesting and informative; it was given with proper informality in the late afternoon, after the heavy science of the day had been concluded; beer and chips were served; and it attracted secretaries, janitors, and book salesmen who happened to be in the building, as well as graduate students and laboratory assistants. The course syllabus (reproduced here with Professor Pead's permission and that of the university) conveys a little of its content, but not much of the pleasure and stimulation afforded to participants.

BIOLOGY 351. GROUP DYNAMICS IN THE IVORY TOWER
A. J. Pead, Instructor

A non-credit, non-required seminar course for graduate students, which emphasizes the management of scientific research groups. Discussion topics include (but are not limited to):

- So you want to hire a technician
- The beauties and evils of graduate student assistants and assistance
- A secretary for all seasons
- The juggling act—grant-supported employment
- Sex in the laboratory: introduction and techniques
- Sex in the laboratory: supervisory challenges and conquests
- Sex in the laboratory: selected case histories and their resolution
- Sex in the laboratory: special problems
- Post-docs, pro and con
- Hello and goodbye—managing staff turnover
- The authorship minefield

- *Morale, discipline and ethics in science*
- *Stress management for science managers*
- *Whistle blowers and whistle blowing*
- *Criteria for identification of good research groups*

Professor Pead, a gregarious outspoken type, had fun with the course, as did the participants, and there was some talk of taping his lectures to be reproduced as a book. His concluding lecture was a classic titled "Criteria for Identification of Good Research Groups." Drawing on case histories which changed annually, he invariably isolated a core of characteristics by which effective groups could be recognized. Some of them that I remember include

- Technicians who are briefed on and are knowledgeable about the rationale and objectives of the research
- Reasonably gracious sharing of after-hours checks on experiments in progress
- Paper cups of champagne at the conclusion of a particularly labor-intensive series of experiments
- Obvious research progress during prolonged and frequent absences of the principal investigator
- A secretary who knows how to wash test tubes and a technician who can type a letter
- A stack of journals on the lab table, waiting to be scanned on lunch hours
- Periodic scheduled informal seminars
- Easy incorporation of visiting investigators into the daily routine.

Professor Pead, recognizing that selection of technicians and research assistants can be a problem for many professionals, also developed a definition of an ideal paid assistant, as "someone with a science background, a near-genius I.Q., training in equipment use and maintenance, computer expertise, typing skills at 120 words per minute, no outside commitments, no psychological hangups, and a perpetually positive outlook on life." He quickly admitted that he had never found such a person, but insisted that the description of an abstract ideal did not prevent him from appreciating, during a long career, some very good and often dedicated assistants.

PUBLICATION PRIORITIES AND AUTHORSHIP

The most important products of much of the labor by the research team are of course publications in professional journals. Few things (if any) are more important to morale of research groups than the ways in which decisions about credit and authorship are handled. These are matters of greater significance—to the graduate student, the technician, the research assistant, the postdoctoral assistant, and the junior colleague—than might be perceived by the principal scientist. Too often, decisions are made too late in the process of data acquisition and presentation, or are made by the principal investigator without full consideration of all the sensitivities involved. Consequences of inadequately examined decisions about authorship can be injured feelings, suppressed hostilities, reduced productivity, or even departure of team members who feel slighted or ignored.

Effective managers of research groups, regardless of the size or composition of those groups, usually recognize the minefield markers surrounding this zone and react prudently. Bumblers can be surprised (even amazed) by the consequences of neglecting common sense rules of passage through such high-risk areas.

An early phase of growing up in science is awareness of the contribution that each team member makes to a final prod-

uct—the published paper—and correct evaluation of the perceptions of the contributors about their relative contributions.

*Recently, members of a multidisciplinary government
research center, geographically dispersed, with a range of
professional and subprofessional staff, were asked to list
guidelines in order of importance that are useful in determining authorship and credit for published documents. The
variability in responses was extreme, opinions were often
forcefully expressed, making it apparent that the subject had
high sensitivity at every level. The initially straightforward
request was quickly transformed (in the best of bureaucratic
style) into a "Delphi" exercise, in which reiterations and
subsets of priority elements were examined and ranked, after
much correspondence and many violent disagreements in
the corridors. The final list of guidelines—admittedly applicable primarily to government research groups and maybe
less so to university or industrial laboratories—contained
the following elements in this order:*

- *Any author whose name appears on a paper must
  understand the paper well enough to present it at a
  scientific meeting.*
- *Multiple authors must not exceed five.*
- *All authors should contribute to the conceptual development of the research on which the paper is based,
  and should participate actively in the research, the
  analyses, and the preparation of drafts.*
- *Multiple authorship is advantageous only to the first
  author; all subsequent authors disappear as part of
  the "et al." in any citation of the paper.*
- *Footnotes or acknowledgments are useful ways to recognize relatively trivial contributions to the paper,*

but are miserable rewards for any substantial contributions.

- *Technicians and research assistants may appear as co-authors, but only if their contributions, in the opinion of the principal investigator or team leader, clearly and substantially exceed those expected as a normal function of the position—and only if they satisfy other criteria for authorship (such as contribution to conceptual development, ability to present results orally, etc.).*

- *Decisions about who is to author the paper, who is to be first author, and what the sequence of authors will be must be made early, preferably before any work is done, and preferably after frank and open discussion among all potential candidates.*

- *The person responsible for the initial concept or idea on which the paper is based must be treated with respect when decisions are being made about who will be listed as first author—provided that that person also contributed to the actual research.*

The furor which developed after distribution of the final list of guidelines was awesome. Opinions occupied a complete spectrum. Every element was challenged by someone— not only for its priority ranking but even for its presence on the list. Technicians felt uncommonly denigrated by several of the elements; junior scientists considered their rights to be unprotected; and principal scientists found their prerogatives compromised. Unhappiness spread to research groups in other locations when copies of the guidelines were disseminated unofficially. The bureaucratic hierarchy was understandably dismayed by the disorder in the ranks which had been created by a well-meaning attempt at standardization.

All copies of the list were ordered shredded, and the exercise
was officially disavowed as "not in the best interests of the
government."

But somewhere, maybe not on this planet, there must exist
an ideal list of guidelines for authorship—one that all members of a research team could believe in and live with comfortably, and one that would reduce the unhappiness which
results from absence of agreement about what constitutes
acceptable practices. Ethics committees of some professional
societies have tried to deal with aspects of authorship problems, but usually only those which border on the unethical.
The larger, more subjective area of justice and equitability is
still a battle zone. Most successful scientists seem to tread very
gently here, insisting on early and frank discussion of authorship with all concerned. The more experienced principal scientists will, however, make it clear that such discussions are
advisory, and will contribute to a fair decision.

MANAGEMENT OF PEER–COLLEAGUE INTERACTIONS

Less critical than intralaboratory relations for the successful professional, but still important to upward mobility, as
many scientists have learned, are cordial relations with peers
and colleagues. Reasonable spontaneity in such relationships
is expected, but elements of planning and management have
been shown to be valuable too. Some of the activities which
can be planned are obvious; others are less so.

Some of the most useful material for this discussion of peer–
colleague interactions was provided by my good friend Dr.
Theodore R. Ross, who has directed a major research labo-
ratory for over 20 years, after an earlier brilliant career as

ing process continues into postdoctoral career stages, with office or seminar discussions with colleagues. Such discussions may even be dragged into luncheon appointments or dinner conversations—to the point of excluding spouses and any nonscientists unfortunate enough to be trapped temporarily in the group, and unassertive enough to stay and tolerate endless "shop talk."

Nonscientists often do not appreciate the excitement levels and the growth potentials of such discussions with sharp informed colleagues.* Interactive groups often speak in an oral shorthand, trying to keep pace with the progress of ideas. Throughout the discussions, thoughts and half-formed opinions are proposed for acceptance, modification, or rejection— to reappear later in more polished or altered form for additional exposure to criticism.

---

*Terms like "peers" and "colleagues" have been used too often in this chapter without definition. Scientists dislike being placed into discrete categories, but they do it to each other regularly. One series of neat labels includes "Authorities," "Peers," "Colleagues," "Other Scientists," and "Scientific Has-beens." These may be defined loosely (and subjectively):

- "Authorities" are highly productive developers of concepts; often older than you, with "credibility" and "status," and with prizes and awards in their recent histories.
- "Peers" are professionals whom you consider to be as good as (or even better than) you are in professional accomplishments, and who are contemporaries in age and position.
- "Colleagues" are professionals whom you consider to be productive members of the scientific community, but not necessarily as good as you are. (They may be younger or older than you.)
- "Other Scientists" are professionals with acceptable (to you) basic qualifications, but who are marginal (in your estimation) in one or more critical aspects of science.
- "Scientific Has-beens" are professionals who have achieved in the past, but whose recent record of publications and other direct contributions to the field has been weak or nonexistent (many scientist-administrators could fit here, even though they may be successful in their new roles).

Conscious or unconscious role-playing can be seen in each group. Some participants with the sharpest minds and the quickest tongues will carry the conversation, while the more analytic ones will be weighing and measuring their contribution for the appropriate entry point. Some will supply the facts and others the occasional comic relief; some make a habit of agreeing and extending, while others almost invariably disagree with whatever is proposed. When the group disbands, most come away with a sense of time well spent; some have their own ideas clarified or thoughts crystallized; some have new insights about old matters; a few remain convinced of the correctness of their original views; and a few frustrated ones go away stamping their feet and shouting obscenities.

The not-so-profound observation here is that most scientists enjoy the company of and discussions with other scientists. The dossiers of successful scientists usually include a notation of cordial long-term relationships with peers and colleagues. A critical element of science is communication of research results to the scientific community; informal as well as formal routes are important.

CONCLUSIONS

Management—of research groups and of relationships with members of the larger scientific community—is an aspect of science that is often underemphasized in graduate education. It is, however, an important ingredient in the ephemeral mix of characteristics and skills which leads to success. Upwardly mobile scientists perceive quickly that individual competence in research—ideas, experimental design, analyses of findings, syntheses—is critical, but that it must be augmented by the varied competencies of many categories of participants in the research adventure, including technicians, graduate students, postdoctoral assistants, junior colleagues, and secretaries. Furthermore, effective relations with members of the larger scientific community are also important in many tangible and intangible ways.

Thus, it seems entirely logical that the first chapter of this book, which dealt with nuts-and-bolts matters (ideas, experimentation, syntheses, communication), should be followed by a chapter concerned with effective management of personal interactions within and outside the laboratory. A combination of these two principal components—individual technical competence and skill in managing people—is, in proper proportions, a worthwhile operational goal for most excellent scientists.

CHAPTER 3

MANAGING CAREER TRANSITIONS

The Delicious Agony Of Major Career Decisions; Satchel Paige Said It All—"Don't Look Back; Something May Be Gaining On You"; Rewriting Career Scripts; Motivations and Guidelines for Career Changes

INTRODUCTION

Dr. Allister Baxter, assistant professor of physics at a New Jersey state college, has a problem. He has just been offered a research job with a small high-tech engineering company, at a salary increase of $12,000. He has two kids in elementary school, a wife with a job selling real estate, and a house with an enormous three-year-old mortgage. The new job would require relocation to a West Coast city; his entire life so far has been spent on the East Coast.

His scientific training tells him to evaluate the matter objectively and rationally, weighing determining factors for and against the move—personal as well as professional. He has done this, but the differences do not seem to be statistically significant. He has asked colleagues who have moved and those who have not moved for advice—with mixed responses. He sees career potentials but also pitfalls in such a radical move. His wife is willing to move, but insists that the final decision be his. Will he go or stay?

The scientific careers of a majority of excellent scientists have been characterized by change. There are, however, transition points where motion is most apparent. Dr. Baxter has come to one such point. These are moments for critical decisions which may shape the entire future course of a career— moments when risk-taking is required, when it is time to break out of established perimeters deliberately, and to invade the uncertain but larger universe beyond. Some responses are based on timidity, some on cautious and exhaustive planning before acting, and some on boldness in the absence of adequate time for planning.

Such moments, agonizing though they may be at times, should be savored fully, especially if they may lead to major career changes or advances. These are fleeting instants when scientists can feel that they are in charge of their destinies, instead of being manipulated in a large electronically controlled game. They are moments of trauma and joy associated with "making it" in science—moments which constitute lifetime "highs" for many good scientists.

Survivors of these high-stress episodes suggest: "once a decision is made, charge ahead with enthusiasm and energy," "once a decision is made, never look back with regret," and "once a decision is made, prove yourself capable of any required metamorphosis in thinking."

Major transitions and upheavals—exciting and sometimes painful—often include the following:

1. Acquisition of the *first professional job,* which marks the final transition from preparation to career, from planning and dreaming to implementing, from theoretical to practical, from starvation to nutritional self-sufficiency.

2. The *great leap forward* occurs with the dramatic acquisition of a job several steps beyond expectations for a given career stage—with selection for a position well beyond present expertise, but one for which existing background provides basic qualifications. An individual's "scope for growth" or capacity for expansion becomes of critical importance once the leap is made. Absolute inner confidence in one's worth and competence as a scientist is a prerequisite for the decision to accept a quantum jump. A surprising number of good scientists lack that confidence, and fall back among the timid and mediocre—where they do not belong, but where they may remain for the rest of their careers. Some important rules of the road for the "great leap" are to take only that part of the past with you that will prove useful in the new job, to make an "all-or-nothing" commitment to the new, and to leave as a departing legacy good relationships with colleagues, since you may encounter (and need) those same people on the declining limb of the growth curve later.

3. The *major career shift*—from teaching to research, from academia to industry, from research to administration—can be the occasion for anxiety unequaled during the rest of a scientific life span.

4. The *midlife crisis* may set in motion entire sequences of changes in a professional career. Some ingredients of dissatisfaction with the status quo include perceived slow progress, disenchantment with current duties or responsibilities (lab sessions, lecturing, committees), and fear of the future. This curious but very real phenomenon will be considered later.

5. The greatest transition for many scientists is often the least obvious—from *scientist to administrator*. This is usually not a single major step, but rather a sequence of minor ones, leading to augmentation of supervisory/management activities and diminution of direct scientific activities. The change can be so gradual that it is imperceptible to the person

involved, until a moment of enlightenment at some intermediate point in the process.

The transition usually involves a gradual accretion of
responsibilities and power—beginning with supervision of lab
assistants, then acquisition of a secretary, then supervision of
other scientists in a group or section, then control of nonscientific components of the organization. A final step is the transition from manager to executive, where effective decision-
making and interpersonal strategies become criteria for
success, and science becomes a fuzzy receding memory. Obso-
lescence of expertise is common, unless active steps are taken
to prevent it—and steps can be taken (such as time dedicated
to literature review or to teaching a graduate course), but at
some personal and career costs, in terms of decrease in leisure
time and less-than-total commitment to new executive
responsibilities.

6. *Sabbaticals* are and should be times of reexamination of
purpose, augmented training in new procedures or thought
processes, introspection, writing, synthesizing, and even for
testing alternate careers.

7. *Retirement* is the final transition of professional life—
not necessarily from activity to inactivity, unless so desired.
Many scientists go on as consultants in their own specialty, or
take up alternate careers in fiction writing, beekeeping, portrait
painting, scientific farming, country inn management—the list
is endless.

An important principle emerges from an examination of
the critical transitions in a professional career. This can be
described as *risk management*—a concept familiar to the business world, but foreign to many scientists. Controllable and
uncontrollable events occur in every scientific life span, and
there is always an element of risk in any transition. The uncontrollable—the chance events—may act for good or harm. On
the dark side, despite careful preparation, elaborate planning,

and statistically sound risk assessment, things may go very wrong at critical points in any career: a valuable mentor may move or retire; a grant application may be rejected without good reason; a key technician may resign soon after the start of a critical series of experiments; or someone else may publish first on a research topic close to your heart. Conversely, random events may happen to boost a career unexpectedly: selection for a position which there did not seem to be a prayer of winning; the technician who decides not to quit after all; or winning a national scientific award with an associated multi-year stipend.

Despite the near certainty of random events occurring, we grope toward models of career-influencing behavior—deterministic models which will reduce the impact of chance events. In part, this is done by careful planning and exhaustive consideration of alternatives. In the end, though, decisions during career transitions must be subjective and internal, after appropriate risk analysis has been made.

REWRITING CAREER SCRIPTS

Examining the careers of scientists on which this book is based leads to the conclusion that many of the successful ones have changed directions—rewritten career scripts—and that the success formula includes at least one major revision during most careers. An additional conclusion is that goals and objectives set early in professional life are often reexamined and modified, sometimes drastically.

Considering the four principal employers of scientists—government, academia, industry, and private foundations—career revisions may involve movement from one to another, at least once, and sometimes more than that. Selected examples of such movements may be useful.

- *Movement from a university position to government science.* Motivation for moves in this direction is often financial. Salaries of government scientists at all levels now exceed those of many colleges and universities. Next in order of motivation is the opportunity to do full-time research in one's specialty—which is particularly attractive to some faculty members at smaller colleges which impose heavy teaching loads and have little organizational tradition in (or interest in) research. Third in order of priority is the availability in many government laboratories of complex and often unique equipment, and the existence of funding to hire assistants and technicians. Lower on the motivational scale, but a real factor for a few, is escape from the stultifying (to some) routine of undergraduate teaching, committee meetings, and faculty politics.

    Scientists may also move from university faculties to full-time government science positions because of the frustrations and anxieties brought on by the perpetual search for grants to continue research—a reality of existence for many faculty members. Another factor is the recent phenomenon of restrictions on tenured positions in universities. This eliminates any job security for those who feel its need (and most scientists are concerned about job security). During the past decade, many good scientists have been recruited to government laboratories from research associate or long-term postdoctoral fellowship positions—all those academic categories with scant job security.
- *Movement from government science to the universities.* Movement from government science to academia is less common than the reverse, but on occasion good productive scientists feel a need for closer contact with students, or become disenchanted with the extensive

document preparation inherent in government employment, or find intractable differences with dictatorial science managers. Sometimes, too, a university establishing or expanding a research program or institute will
urge good government scientists to accept key positions.
Fringe benefits and salaries in government are strong
disincentives for change, however, unless early retirement is feasible before the move from government.

---

*Dr. Boris Chapayev is an example of the younger (early for-
ties), highly trained, management-oriented government scientist cum administrator. In a period of less than a decade
since receiving his Ph.D. he has published substantive
papers in quantitative science, he has supervised progressively larger research groups, and he was picked three years
ago from a number of good candidates to take a quantum
jump to become director of a research complex with laboratories in several states. For the new position he sensibly
applied himself to his managerial activities, and did not try
to carry too much research baggage on the trip—preferring
to delegate research supervision to his laboratory and division directors.*

*As a research director, he acquired a reputation as a
knowledgeable, hard-driving, no-nonsense administrator,
aggressive in funding matters, as all good government managers must be, but also a reasonable team player. To every
observer he clearly belonged at the executive/policy level of
the agency, and seemed to be headed in that direction.*

*But then, surprisingly, he resigned to accept a univer-
sity offer—to head a new and growing institute, at a substantial reduction in salary. The stated reasons for the
abrupt change were the usual ones—disenchantment with
federal bureaucracy and a desire for closer academic associations. Privately, Dr. Chapayev confided that the new insti-*

tute directorship was an irresistible challenge to him, since it offered an opportunity to find out what his capabilities were in fund-raising, political interactions, and research planning within a university structure. Also, the new job offered him a chance to make contributions to policy about graduate education in this country—a subject of great personal importance to him. Furthermore, he admitted, he and his wife liked the climate and life-style of the city in which the institute was located.

- *Movement from universities to industry.* Some universities have extremely liberal policies about consulting activities by their faculty; this can mean significant augmentation of marginal academic salaries, and retention of some faculty with higher financial aspirations. Once in a while, though, an entrepreneurial faculty member may decide to go the whole route and form his or her own consulting firm or technology-based company—alone or with other faculty members. This giant step is a high-risk move; a few succeed in this calling but many don't. Those who do succeed often recruit university colleagues to act as consultants, and hire new Ph.D.'s or students from the university, frequently on a part-time or temporary basis.

  For those scientists with credibility, expertise, and drive, financial returns from private ventures can be substantial, although the price is usually cessation of active research in favor of management responsibilities, and extensive travel in search of new contracts or customers.

- *Movement from government to industry.* Several categories of scientists make the move from government to industry. The emergence, especially during the past two decades, of technology-based companies has resulted in recruitment, at high salaries, of many bright, aggressive junior and midlevel government scientists, bringing with them very specialized expertise. This phenomenon has occurred in small entrepreneurial companies, or in new subsidiaries of well-established diversified business organizations.

Dr. Dana Brawner headed a small specialized research group of four other professionals and assorted support staff. Joining the federal laboratory of which the group was a part 12 years ago in a relatively junior position, he prospered professionally inside and outside the organization. Largely as a consequence of good and extensive research, and its publication in major journals, he achieved national recognition as one of the leaders in his specialty, which is undergoing rapid expansion.

Dr. Brawner has been an invited participant in a number of workshops, symposia, and conferences, where his research results and his opinions have been highly regarded. He has served as a visiting lecturer in several major universities, and he declined offers of faculty appointments in favor of the research he was doing. He is charismatic, somewhat dictatorial, and a decided intellectual snob, but a good scientist.

Two years ago he was approached by a group of scientific entrepreneurs with specialties similar to his, and was invited to join their newly formed high-tech company—with stock options in place of a large salary. After several agonizing months he accepted the offer and resigned from his government job. The new job has turned out to be mostly promotional, with extensive travel, and little hands-on research—but (on paper) he is getting rich and he claims to be happy in his new role.

Another important channel has been departure of highly trained productive government scientists through early retirement—and their reappearance in key positions in industrial organizations of all types, but particularly in defense industries. Sometimes these early retirees, alone or in groups, will form consulting companies, with government agencies as their principal customers. In other instances the retirees join larger companies but act as contact points with their former agencies.

MOTIVATIONS FOR CAREER CHANGES

Underlying reasons for career changes include desire for professional advancement, interest in working on more stimulating projects, need for higher salary, expectation of greater recognition for accomplishments, and a host of others in descending order of importance. Interest in a career change is translated into action by a long list of precipitating factors, variable with time and with individuals. Review of the scientific life histories which form the wobbly statistical base of this book discloses four principal precipitating factors—financial crisis, midlife crisis, divorce, and death of someone close.

Money, or its absence, looms large in many early scientific careers. Graduate school is expectedly poverty-ridden; most grants and fellowships are designed to permit survival but not well-being. Salaries attached to introductory-level jobs in science improved remarkably in the 1970s and early 1980s, but they still lag behind expectations in other professions—particularly in the small private colleges, which even today may have pay scales reminiscent of the 1950s. Most universities and the federal government have done more to upgrade entry-level science paychecks, but in the absence of a hard-working spouse or an inheritance, the financial basis for an early career in science is often precarious.

All this financial uncertainty raises the priority level of earned dollars, and increases the likelihood that a career decision will be based on money. Precipitating events vary to the extreme, but could include a new baby, demise of a car, need for a bigger house, a 35th (40th, 45th) birthday accompanied by an assessment of present net worth, nursing home bills for aging parents, the prospects of college tuition bills only a few years away—or an almost infinite number of comparable financial pressures. Career potentials must be considered in any job change, but money is a large factor in many, if not most, decisions.

Beyond the never-ending need for money, and its influence on job changes, some scientists are subject to acute attacks of what has aptly been termed "midlife crisis." A day arrives when they look in vain for conceptual advances or brilliant syntheses that they have been instrumental in developing—when they realize the transiency of their publications, even within their lifetimes, and when they seem surrounded by aggressive, highly qualified juniors at faculty or staff meetings.

The midlife crisis has a higher statistical prevalence among middle-aged men, but is not infrequent among women of comparable age. It is a stormy time of complex feelings of unfulfillment, of retreat to immature judgments and actions, and of a vague search for greater meaning in life. It is a time of mindless, often juvenile, behavioral changes and reversals—a time of regression, depression, divorce, alcoholism, and suicide. Successful people are as susceptible as the not-so-successful. The successful perceive themselves as failures, and the unsuccessful are often overwhelmed by feelings of self-pity and self-hate.

It is entirely logical that some of the career revisions done by scientists should evolve from internal conflict and disorder during this period of awesome psychological malaise. The wonder is that more poor decisions are not made, since judgment is clearly impaired. The explanation must be that the period of acute disturbance is of limited duration, usually lasting no more than a few years—not always long enough to destroy totally a heretofore productive, even successful, career.

Other forms of personal trauma can be involved in or can precipitate career revisions by scientists. Next to midlife crisis, divorce is probably the most frequently observed factor in a decision to revise (it may be statistically the most important single factor for female scientists). This phenomenon is of course not unique to scientists; a common postdivorce activity is relocation in a new job somewhere else. Unlike some easily transferable occupations, though, scientific job changes usually require adequate lead time and advance planning.

Death of a family member, especially a spouse or child, can be still another precipitating factor for career revision. The desire for escape from familiar scenes and faces can be particularly strong at such a time.

GUIDELINES FOR CAREER CHANGES

Considering the extreme diversity of job changes and the varying circumstances for each one, it is difficult to find common threads to develop a pattern of rational behavior at such times. Having spoken with many survivors of these critical periods, however, I find that a few general guidelines do emerge:

- *Timing.* There are correct and incorrect times to revise. The best time is when an attractive option is presented. The worst time is when the job market is abnormally tight.

- *Planning.* Major revision takes careful planning and preparation, often extending over months or years. It should not be haphazard or precipitous, unless extraneous factors force a decision.
- *Justification.* Reasons for revision must be sound unless a change in jobs is dictated by outside forces (this may be the sole reason). Obviously it is better to revise voluntarily than to have a revision dictated by others.
- *The short-term flirtation.* A subset of career revision which seems in vogue at present can be labeled "temporary aberrations," or "short-term flirtations" with other kinds of jobs. These brief encounters of an exploratory kind have been made easier recently by a variety of government-sponsored plans whereby university faculty members can take short-term appointments with government agencies, or where government scientists can join university teaching or research staffs for equally short periods (usually less than two years). Sabbaticals can of course provide similar short-term professional experiences in different jobs.

  Advantages of these flirtations are that new career directions can be explored at little risk of unemployment, and without long-term commitment. New perspectives will certainly be acquired. The principal disadvantages are a loss of position in the internal pecking order of the parent organization, and the creation of a perception within that organization that the temporary sojourn elsewhere may be an indication of unhappiness with the original environment.

The recent exploits of Dr. Anne Merriweather provide us with an excellent example of full use of the "short-term flirtation" concept. The university at which she was a highly respected faculty member was undergoing the twin traumas of a change in administration and severe financial restraints. She requested and was granted a two-year assignment under a government program (called the Intergovernmental Personnel Act), spending the first year with a government granting agency and the second with a research and services agency, doing mostly staff work. During that two-year period she occupied a number of staff positions at several levels, sat in on innumerable policy, planning, and evaluation meetings, and developed many positive contacts with government scientists and bureaucrats. She then returned to the university, which in the interim had achieved some stability, and resumed her research and teaching, with a far better appreciation for, but no greater attraction to or affection for, the processes of government which affect science.

The positive consequences of her bureaucratic fling, in her opinion, were reaffirmation of the rightness of her choice of a career in teaching and research, and acquisition of a broader perspective on the national importance of scientific research in the United States.

- *Career plateaus: Causes, cures, consequences.* An important but underestimated motivation for change can be the realization that a "career plateau" has been reached. Such a plateau can be defined as "any period of five years without significant advancement in salary or status; any period of two years without a substantive publication (beyond the potboiler category); any period of a year without presentation of at least one scientific paper at a scientific meeting; any period of a year without an invitation to present a seminar or lecture at a location other than the workplace; or any period of six months without substantial numbers of hours of direct research involvement."

  Causes of career plateaus may be obvious or

obscure, including lack of job niches at higher levels of the organization, persistent mediocrity, timidity, ineffective participation in organizational politics, existence of an enemy in the approval chain for promotion, or chance events which block progress.

Overlong tenure in a comfortable rut may be pleasant, but the productive life span of a scientist rarely exceeds 30 years, and options begin to narrow if some form of recognition is not achieved during the first 10 years of professional life. The fixed sequences of a career may be delayed progressively, so the expected peaks may never be reached, without a serious game of catch-up. The five-year maximum allowable plateau in a career is of course debatable, but it is also defensible, in examining the careers of successful scientists. A similar maximum applies to other types of jobs; whatever impact an individual can make on a job will usually be made within that period, and then it is time to move, unless future rewards for staying are clearly visible.

At any rate, recognition of the existence of a career plateau calls for aggressive actions, which may include (but are certainly not limited to) deliberate involvement in a research project which will require long hours every day, including weekends; personal discussions with one's supervisor or department head; or frank discussions about the future with one's spouse.

- *The tenure trap: A major deterrent to change.* In examining the careers of scientists, it is remarkable how frequently security, in the form of tenure and organizational retirement plans, hinders mobility. The script is dictated by the system; future career planning is governed by past events, and the limits to a career are imposed by the location of the job. Seen from this perspective, nontrans-

ferable tenure and retirement funds account for much of
the perceived stability, except at the top and the bottom
of any success scale. It is the middle-level people—those
clustering around the norm in ability, competence,
energy, and perception—who are trapped by the system. The top of the scale includes those professionals
who are exceptional, who are risk-takers, and who
always have employment options. The bottom of the
scale includes those marginal professionals who have
found a comfortable berth and cling to it desperately, in
the absence of options elsewhere.

Looking at these guidelines for career changes dispassionately, it is obvious that a mixture of objective criteria and subjective factors will always exist, and that decisions will be, in
the end, based on intensely personal elements.

CONCLUSIONS

Analysis of the careers of large numbers of successful scientists leads to a number of reasonably founded generalizations about managing transitions:

- Major career revisions are the *norm* for many successful
  scientists. This is the *usual* route to success, rather than
  continuation in a path selected early.
- Revisions often accompany some form of personal
  agony. Disruptions in personal life—financial problems,
  divorce, death of a spouse or child—can be precipitating
  or contributing factors.
- A factor which is also associated with revision is the
  well-named "midlife crisis"—predominantly a feature
  in careers of males, but not unknown among females.
- Good scientists seek new challenges. Some can be found
  in extensions of existing jobs, but some must be reached

for outside the perimeter of the familiar and the expected.

- Revisionists form a high proportion of the successful segment of the scientific community. It may be that this segment of the population contains the calculated risk takers—and that they are able to take risks because of internal feelings of worth and ability. It is probable, too, that in this segment are the demonstrated producers in some aspect of science, who are offered other broader options.

Some professionals set a career course early and maintain it, whereas others reach decision points at various stages in their lives which lead to major shifts in the kind of science they practice—and even on rare occasions to total disappearance from science.

PART TWO

SCIENTIFIC ELITE

Excellent scientists share at times, with good people in other fields, feelings of worth, control, and accomplishment. Their career journeys include numerous peaks of major publication, significant insights, successful syntheses, and recognition from peers. This second major section of the book examines, analyzes, highlights, and otherwise tries to grapple with concepts of success and excellence in science as a discipline and occupation. It concentrates almost exclusively on those who exemplify such positive characteristics, in an attempt to answer questions about "why" and "how."

Treatment here of the elite of science begins haltingly and a little pedantically with a discussion of the external indicators of success, then flits quickly to the organization of successful science, emphasizing the "club" and "networking." Once the externals are accounted for and stripped away, the much more complex interior of how scientists approach their tasks is exposed. The chapter on "Internal Journeys" wanders around in this analytic combat zone briefly and with only marginal effectiveness. Retreating prudently to firmer ground, the chapter on "Destinations" scrutinizes some of the kinds of jobs that

successful scientists perform—many of them actually well outside the research and teaching stereotype. A chapter on female
scientists—their problems and successes—follows, for no logical reason except maybe to propose some half-truths (or
untruths) about male–female interactions in the scientific
workplace. After this catharsis, the section draws to a dramatic
close with a chapter on "Authority Figures," emphasizing the
tenuousness of management-based authority and the relative
permanence of science-based authority.

CHAPTER 4

EXTERNAL SIGNS OF SUCCESS

Recognizing the Successful Ones—Scientists Who Make It and How They Do It; A Preliminary Scuffle with the Terms "Success," "Excellence," and "Good Science"; How the Success Game is Organized—Clubs, Fraternities, and Networks; Scientific Elite and Elitism

INTRODUCTION

In every technologically advanced country in the world there are professionals—some working at night in small isolated laboratories, some ensconced comfortably with arrays of assistants on large university campuses, some working with small groups of colleagues in industrial laboratories—all confident of making significant contributions to understanding in their own narrow half-acre of expertise. Some make it, as measured by prestige and significant scientific contributions; many don't. The competition is intense, and the external satisfactions are often few.

What is it that separates the routine from the exceptional, the great from the almost-great? What spark or internal fire forces a professional to make that extra commitment of time, energy, and mental effort which may eventually result in all the indications that he or she has succeeded in science? As usual, the answer is complex and unsatisfying.

Scientists rarely set out to achieve "success" as an abstract or stated career goal (at least insofar as they are willing to admit). Their stated objectives are more likely "to contribute in a significant way to knowledge in a chosen field," "to do the meaningful things in science that I enjoy most," or even in a few cases "to make a living." However, in attempting to fulfill stated career objectives, whatever they may be, some scientists do achieve what is viewed by colleagues as "success." Examination of the routes that they follow discloses some commonality of accomplishments. Included are some or all of the following:

- Substantial original research in an area of expertise
- Extensive publication in the best journals in that discipline area
- Publication of substantive, authoritative, superbly written reviews and technical books
- Significant contributions to ideas, concepts, analyses, and syntheses in an area of expertise
- Effective participation in workshops and symposia
- Effective oral presentations in society meetings and at conferences
- Ability to attract good graduate students and to aid substantively in their transformation into scientists
- Development and maintenance of close professional and personal relationships with a group of productive colleagues

- An almost instinctive ability to assume a dominant role in committees, informal workshops, and similar group activities (this stems in part from credibility as a scientist, enthusiasm for the subject matter, and careful consideration of the prerogatives of others)

The word "success" must be used with great delicacy, since it has so many connotations, nuances, and interpretations. It is a profoundly personal term, for often what is viewed by colleagues as an example of success in science may be viewed internally by the individual exemplar as mediocrity or failure. Criteria of success (such as money, fame, or advancement in rank) important to one professional may be less important to another professional. Personal perspective thus shapes and modifies the meaning of the term.

Despite these variations in interpretation, though, there does seem to be a core of indicators which would be accepted as characteristics of the more successful scientists as a group. These include some or all of the following:

- Achievement of respect from colleagues and "credibility" among them as a thinker and producer
- Acceptance into and active participation in the real but ill-defined "in-group" which exists within any sub-discipline
- Nomination and election to society offices, and longterm participation in society boards of directors or standing committees
- Frequent and persistent invitations to participate in symposia, to chair sessions, and to present lectures or keynote addresses
- Occasional invitations or nominations to chair national or international meetings, workshops, working groups, conferences, and symposia

- Invitations to serve on editorial boards of journals, and (on rare occasions) even to act as editor of a scientific journal
- Appointment to research administrative posts as group leader, division manager, or laboratory director

This core of indicators may characterize successful scientists as a group, but there are numerous individuals who produce excellent and substantial scientific findings who would satisfy none of the above criteria and who would reject every one of them as superficial and nonessential. To such professionals, success must be defined and measured in the narrower technical sense, by hard objective assessment of quality and quantity of scientific productivity, unclouded by any interpersonal "trivia." This "no-nonsense" approach to science must be given its due. Certainly scientific productivity should never be reduced to insignificance in any kind of assessment—but productivity in the absence of participation in the larger world of science seems somehow less than ideal to many excellent scientists with whom this matter was discussed. They point out that the development of science is based on exchange of ideas and findings with others, not only through the printed word, but also through discussion. But they point out further that even the humblest, most reclusive scientist whose work is of excellent quality is likely to be recognized by colleagues.

The case histories described in this book tend to emphasize (perhaps unfairly) the big winners—the charismatic leaders of organizations or areas of research—but there is a much larger corps of professionals who consider themselves to be successful, but who meet few of the criteria proposed in this chapter. Members of this group have their own standards of success, which are often quite different from general criteria. These standards may include completion of a unit of work as a unique contribution to scientific methodology, analysis of a data set in an original way, the formulation and initial validation of a minor concept, or preparation of a significant review in a specialty area. To many such scientists, "success in a minor key" is a worthy career objective, far from the glitter of national advisory committees or international symposia.

The reality usually turns out to be that each professional elects to make a maximum investment in some aspect of science. Some make it in hands-on-bench-level research, some in writing, some in supervision or administration, some in editing, and some in synthesis and concept development. Whatever the principal commitment may be, other aspects suffer. Criteria for success, therefore, cannot follow precisely the list of indicators given above, since to excel in every direction is rare indeed.

DEFINING "SUCCESS"

The preceding discussion of "criteria" and "indicators" of success still leaves a vague feeling of unease—that we have not grappled effectively with the *concept*. The words "success," "succeeding," and "successful" have already been used too many times in this book and we are only as far as Chapter 4, so it is obvious that some more strenuous attempt to define "success" is required. The concept seems superficially simple, but its codification is not simple, and its essence is surprisingly elusive. We must first of all distinguish as clearly as possible between success in *doing* good science and success in the many *interpersonal activities* which surround the process. The two aspects are often but not always related, and are often but not always embodied in the same persons. Good scientists frequently have great rapport and empathy with peers and colleagues, with laboratory assistants and graduate students, and with administrators or other scientific bureaucrats. Sometimes, though, good scientists can be impossible as human beings, difficult to live with as colleagues, and deliberately ignorant of the rules and niceties of almost every kind of social interaction. Certainly it must be possible for professionals in the latter category to be considered successful if their scientific productivity alone warrants it. The converse is not true, however; professionals adept at interacting with people cannot be considered successful *as scientists* without the requisite technical productivity. They may be successful as administrators or teachers or politicians (all to be considered in great detail in later chapters), but they would not be regarded as successful *scientists*.

Another critical aspect of any attempted codification of "success" in science concerns *perceptions of success*, particularly by the individual professional. Internal goals and internal feelings of worth and achievement are probably more important in science than in many other occupations. This does not imply that goals and rewards imposed externally by colleagues and society are unimportant, but it does suggest a broader dimension to any assessment of success—one in which the scientist's own evaluation can be of overriding importance.

Success in technical aspects of science is important, and for some it is the only goal. For many scientists, though, that kind of success must be accompanied by good relationships with and the acclaim of colleagues, if full pleasure is to be attained. Some of the satisfaction and joy is clearly internal— in knowing that something meaningful has been accomplished—but the pleasure is often greater when achievement is recognized and acknowledged by others. For many of us, there must be someone out there, qualified to judge, who accepts and approves and occasionally applauds. That someone may be a colleague, a supervisor, a laboratory director, or a dean. The approval should be genuine, not perfunctory, and it must be earned to be of value.

DEFINING "GOOD SCIENCE"

Attempts to define terms which have no precise definition can be like claiming Fifth Amendment privileges in a court-room—once that route has been elected there can be no hiding place and no acceptable retreat—the action must be carried with great consistency, regardless of consequences. The term "good science" has already been used several times, and since it will be used throughout this book, it might be worthwhile to try to define it and to describe some of its connotations. Any single definition would never satisfy even a significant minority of practitioners, so the attempt is futile from the start—but this should be no deterrent.

"Science" has many definitions; most of them include the concept of verification of natural laws through specific observations—attempts to understand the physical, chemical, and biological universe—attempts which are progressive but always short of perfection. Following this conceptual sequence, "good science" could be defined as those practices which contribute most to advances in understanding, but even "good science" as a general term encompasses a spectrum from good to excellent:

"Good" *science* might be said to apply to sustained research productivity in a chosen area of subject matter, supported by substantive papers in major journals.

"Very good" *science* could connote a series of major research papers extending over a number of years, which in the aggregate provide a substantial addition to knowledge in an area of subject matter, or definitive and creative reviews in a specific subject, or specialized books in an area of expertise.

"Excellent" *science* could include a brilliant series of definitive research papers, exploring in depth a previously little-known phenomenon; a significant conceptual advance; a masterful elegant synthesis of the disparate data of others; or a

definitive, award-winning basic textbook in a major scientific field.

HOW THE SUCCESS GAME IS ORGANIZED

We have been delayed too long by definitions and descriptions of "success," "good science," and similar upbeat terms; it is now time to explore the reality of those elusive concepts, building on the distressingly weak infrastructure created thus far. To do this, the approach chosen is to consider first the *outward signs* of success in this chapter, then to examine the more important *internal components* of success in the next chapter.

Analyses of the case histories of successful scientists disclose numerous common themes, many of which can be prodded and extrapolated into useful generalizations. Some are more obvious than others; some are probably misleading; but some can provide respectable insights into the nature of success and excellence in science.

Organization of the success game is loose but real. Some of the internal framework can be seen at any national society meeting or symposium. An admittedly subjective analysis of participants discloses the presence of four reasonably discrete categories:

- The *"club"*—the real "inside group" in any scientific society or subdiscipline of science
- The *"fraternity"*—a junior version of the "club" consisting mostly of bright vocal assistant professors and post-doctoral fellows
- The *"outliers"*—scientists with only marginal or peripheral involvement in events of the day
- The *"drop-ins"*—a motley assemblage of administrators, bureaucrats, consultants, and undergraduate students

The first two categories deserve close attention here, since members of those groups often fit our various criteria for success, and they usually participate in communication networks which have proved vital to their success.

THE "CLUB"

In any subdiscipline of science there exists a small select informal "in-group" that can best be called the "club." Its members are good thinkers and are politically astute. They may be but are not necessarily the best in the field (in terms of original research contributions) but they are the ones who seem to participate most consistently in the decisions which affect science. They are the ones invited to policy-level workshops and conferences sponsored by the National Academy of Sciences or other prestigious bodies. Club members present "overviews" at national and international meetings; they write reviews; and they chair advisory committees. Election to and continuing membership in such clubs is an excellent sign of "making it" in science, although good people are often excluded—sometimes for petty vindictive reasons, or because it is mandatory that the group not get too large. (Informal limitation of membership in the club is an expansion of the "Forty-first Chair" concept developed by Robert K. Merton in his journal article "The Matthew Effect in Science."[1] The concept stems from a practice of the French Academy of limiting membership to 40, thereby excluding in any generation and over the centuries many exceptional and deserving individuals—some of whom, for example Descartes, Rousseau, and Proust, went on to achieve immortality without the honor of membership in the Academy.)

Depending on the subdiscipline involved, some clubs are more rigid and exclusive than others. In areas dependent on public funding for research (a majority in today's climate), club members play principal roles in determining the direction and size of grant pipelines (often through membership on permanent advisory committees which recommend policies to funding agencies). Club members also control the selection of the best graduate students, through informal but constant communication networks, and help to determine their destinies, by including students in their own active research groups or by recommending them for placement in choice jobs.

THE "FRATERNITY"

A separate subclass with priority ranking just below the "club" is the "fraternity"—a group of extraordinarily bright, well-trained, aggressive postdoctoral people mostly in their early thirties. They are often the brightest protégés of members of the club, and they are usually in junior or intermediate faculty positions. Fraternity members* relate so well to one another that they speak in telegraphic jargon phrases, effectively excluding others not involved in their narrow, highly specialized subdiscipline. Fraternity members are highly mobile, appearing at many national and international meetings and traveling to foreign countries, often for extended periods of research. They tend to be extremely cohesive, and somewhat condescending to "ordinary" scientists. They also have the pronounced handicap of extremely limited perception of the world, whose perimeters do not often extend beyond those

---

*The "fraternity" label may be somewhat of a misnomer today, because of the entry of more and more females into this junior scientific chamber. No exclusive sororities exist yet, to my knowledge, but the bars are down in the scientific fraternities. That increasingly inappropriate name is still the best descriptor we have, however.

of a narrow subdiscipline—to the point where their entire verbal existence focuses on it. The fraternity is the farm team for the club.

"NETWORKING"

The development of a mutually supportive, interactive, ever-widening circle of those with similar objectives, abilities, and views—called "networking"—has received recent publicity as a particular consequence of the women's rights movement. It has for a long time been a principal operating device of politicians at local and national levels. Moreover, the concept, not always formalized by a label, has been employed by successful business executives, Army supply sergeants, and used-car dealers.

Successful perceptive scientists, especially those in managerial roles, have made extensive use of the concept. Its essence consists of *constant substantive communication*. It begins with a selected group of colleagues; its limit is only that of time for phone calls, letters, memos, and corridor discussions at scientific meetings. Some principal elements of science networking are

- Regular and frequent discussions (phone, letter, telex, etc.) with peers and colleagues, whether the subject matter of the moment is critical or not
- Inclusion of those same peers and colleagues in planning for workshops, symposia, or conferences
- Frequent late-evening small-group conferences in hotel rooms, or poolside discussions during scientific meetings
- Requests to peers and colleagues for preliminary review and comments on manuscripts, position papers, propos-

als, and a host of other documents—followed by use of the comments and suggestions received

- Requests to peers and colleagues for suggestions or comments about research projects and interim reports
- Reciprocal acceptance of peers' choices of excellent students for graduate training
- Immediate acceptance of requests from peers for participation in workshops, symposia, or sessions which they are organizing, followed by effective enthusiastic participation in those activities

Skill in networking can lead quickly to additional activities on a higher level—properly called "advanced networking." A sampler of ingredients includes

- Periodic authorship and distribution, in multiple copies, of circular letters on scientific policies or issues of the day; or even
- Developing, editing, and distributing a newsletter for members of a small core group in a scientific specialty (soliciting contributions from other members of the group, but retaining editorial control)
- Organizing workshops, working groups, task forces, and committees to fill a recognized need—retaining the position of convenor, chairperson, or rapporteur
- Organizing a summer institute or an informal "invitation only" extended conference of key people in a scientific specialty; or even
- Organizing an institute or foundation to perform specific needed functions in a scientific specialty

The extrapolations of "advanced networking" are limited only by imagination and time, but the key is always the same—increased communication with peers and colleagues.

Dr. Georgine Sunderland, wife of my good colleague J. C. Sunderland and a practicing psychologist, once gave, at our request, a seminar series titled "Networking for Scientists," which was later remodeled as a paid one-day course for scientists, engineers, and administrators. Many of the preceding points were covered in her seminars, but she also listed some principles of networking that extended beyond the "how to." They are considered here with her permission, and include

- *Networking is a <u>mutual</u> activity in which something is given as well as received. Without this reciprocal exchange there is no true network.*
- *Networking is a <u>controllable</u> professional activity, in that the individual determines the extent and nature of involvement. Thus, networks may be expanded as personal interest in a particular subject or research group increases, or contracted as personal interest diminishes.*
- *Networks can and should be <u>purged</u> periodically, to weed out the noncontributors, the bores, the time-monopolizers, the self-servers, the nonresponsive, the overly demanding, the impotent. Expurgation may be a natural consequence of changing interests, or it may be a planned annual event. Pragmatic decisions must be made about casting out nonproductive contacts, keeping some marginal but pleasant contacts as pets, or keeping some as insurance against future needs.*
- *Conversely, some networks need to be <u>reinforced</u> and revitalized periodically, since they are dynamic, and can suffer from disuse. Members should get a maintenance phone call, memo, or note every month.*

- *Networking is multidimensional, functioning simultaneously on many planes. From the perspective of the individual, he or she represents the intersection of many of these planes, each of which may be peopled by different kinds of scientists, and each of which may require different kinds of responses.*
- *If the plane analogy is not descriptive enough, each network can be envisioned as a classical elementary textbook drawing of atomic structure, with a nucleus or core and with expanding rings of electrons as satellites. The core group is preserved and nurtured with great vigor; the satellites can be held in looser and looser orbits of decreasing relevance.*
- *Networks constitute a form of public release of information, opinions, and ideas. Members of any network should be sensitive to this in determining what information to offer to which participants.*
- *Networking has a major social component. To many scientists, job-related contacts are important in a social sense; some of the most intimate and supportive personal relationships are with colleagues—more so than is found in other classes of professionals (physicians, lawyers, dentists, and certain others).*
- *Networking frequently implies making commitments—and carrying through on those commitments, providing assistance when promised.*
- *Every useful new contact, regardless of how fleeting, should be exploited. Follow-up phone calls, expressions of genuine appreciation for comments (if warranted), requests for further information—all are natural networking necessities.*
- *Asking a favor is often a good way to open access—provided that there has been a previous introduction or that a common link is obvious.*

- *A copying machine is the networker's best friend, enabling expansion of the active net from dozens to hundreds.*

All of this networking discussion leads to a final mildly overpowering insight: a major ingredient of the jobs of most scientist-administrators and many scientist-bureaucrats is the development, maintenance, and exploitation of a complex network of contacts in other organizations, local, national, and foreign. Much of the typical workday is spent telephoning, telexing, or otherwise communicating with individuals throughout the network. Constructing and upgrading the connections are not simple tasks and not easy to carry out. Sometimes years of effort are required, and it seems that there are always additional links needed. The construction can be a principal career objective, since effective job performance (and often survival) can be dependent on skill at building networks.

This insight occurred to me with great clarity while listening to an agricultural specialist describe his job in the upper echelons of an international development agency (AID). He described links with other specialists in the United States and foreign countries, and their links in turn with other networks in other organizations—all being essential to his ability to carry out assignments which involved cooperative projects, international assistance programs, and sensitivity to emerging crises. Probing further, I found the same thing true of scientist-administrators in other agencies, such as the NSF, NIH, and USDA—all agencies with granting or contracting functions.

SCIENTIFIC ELITE AND ELITISM

The elitist bias of this book, acknowledged in the Prologue, can be a defect or an advantage, depending on points of view. Successful scientists constitute an elite group from any point of view, however, and it seems entirely proper to accept this as fact—whether the selective process is based on productivity, brilliant insights, or interpersonal skills. Such acceptance leads quickly, though, to disquieting questions such as "What, then, is the *average* scientist?" or "Where does scientific snobbery begin?" or "Is success in science a self-reinforcing process (in the sense that those who have achieved go on to further achievements)?" These musings, fortunately, have no clear resolution. "Average" scientists exist mostly in perceptions of colleagues and not in self-images; snobbery in science or elsewhere is also a perception of those not included in any "in-group"; and there is evidence for concluding that those who have achieved find it easier to achieve again in subsequent work.

In the interest of getting on with this too-brief consideration of elitism, we can begin simplistically with a listing of a few common distinguishing traits of the scientific elite:

- *They frequently possess liberal quantities of whatever is implied in the overworked term "charisma."* The personal characteristics which induce positive responses from others can be important to success in the interpersonal components of science. Scientists defy characterization by physical traits—some are short, tall, lean, plump, disheveled, preppy—but many project the image of interesting concerned human beings.
- They know internally that *they are professionals who are good at what they do;* they always *act* and *think* as professionals; they expect to be treated as professionals; and they are always "on duty" and "on stage." Furthermore, they expect professionalism and excellence from colleagues and subordinates—and often get them.

- Many scientists are only marginally social beings, so they can be very uncomfortable when thrust on the public scene or expected to interact effectively with numbers of people. *Successful scientists, though, are those who adapt*—who learn interpersonal skills or who have native abilities which can be exploited when dealing with colleagues or the public.
- *Successful scientists rarely function professionally as isolated individuals.* They tend to create and to surround themselves with an effective loyal support infrastructure consisting of mentors, assistants, technicians, secretaries, postdoctoral fellows and aides.
- They are firm believers in and subscribers to the "Matthew Effect" in science, as elucidated by Robert K. Merton in his article on that subject,[1] which states (in general terms) that *the reward system in science favors those who have already achieved.*
- *They are never satisfied with a "routine eight-hour day."* Days are never "routine" and good science does not necessarily fit the normal business day.

CONCLUSIONS

"Success" in science has too many nuances and connotations to be defined with any great precision. A basic component of any description, however, must be productivity—significant contributions to knowledge about natural processes. Without this base all other possible components have little meaning. Once the criterion of productivity has been satisfied, however, some of the more social aspects of science can become important for many practitioners. Included would be close professional relationships with colleagues, participation in activities of professional societies, service in reviewing or editing for journals, and organizing or chairing workshops. External signs of success in science are thus a blend of technical accomplishment and skill in interpersonal relations.

Looking more closely at the organization of successful science, it is easy to discern the leaders in any subdiscipline—loosely aggregated into "clubs" and "fraternities," and communicating with each other through elaborate informal networks. Membership and active participation in this scientific infrastructure can be excellent indicators of "arrival" in science, and can contribute substantially to the pleasures of an exceptional occupation.

REFERENCE

1. Robert K. Merton, The Matthew effect in science, *Science* 159 (1968), pp. 56–63.

CHAPTER 5

"INTERNAL JOURNEYS" OF SCIENTISTS

What and How Scientists Think; Internal Foundations of Scientific Excellence; The Nature of Insights; Joys of Science

INTRODUCTION

Thus far in this book we have been dealing principally with externals—with obvious and somewhat superficial criteria of success in science. It should be apparent that we have not yet penetrated to the thought processes of scientists—to which producing, publishing, interacting with peers, and worldly rewards are all secondary. According to a recent biography *The Science and the Life of Albert Einstein* by Abraham Pais,[1] Einstein once wrote that "The essential in the being of a man of my type lies precisely in *what* he thinks and *how* he thinks, not in what he does or suffers." To those of us who would understand the meaning of "success" and "excellence" in science, this statement should be considered with great care. It implies that in any serious examination we would do well to consider basic thinking patterns as well as relative superficialities of producing and publishing (even though the superficialities are more accessible).

What scientists think is founded to a great extent on the thinking and the work of others who have developed concepts, have acquired supporting data, and have published their findings. These predecessors have "cleared the trail" and "set the markers" up to the present frontier. They have determined the existing boundaries between ignorance and knowledge, and they have often provided clues about future research directions.

Science is an art form. Just as with painting or sculpture there are those who develop its commercial aspects, using "formulas" to produce a saleable product. Some professionals stumble on a formula early and grind out successive papers which elaborate on a single theme. Then there are those who search for elegant portrayals of concepts—of visions—and who become bored quickly with the routine and the mundane. Such individuals may spend an entire career searching for but never finding the right experiment or never making the unique synthesis. It is from the ranks of those searchers, though, that brilliant new insights can be expected.

How scientists think (as contrasted with *what* they think) is more difficult to comprehend, but is a worthwhile subject for analysis. They must use *inductive and deductive reasoning,* vibrating from the specific to the general and back again repeatedly; they must be *perceptive,* able to recognize a critical component or new fact when it appears; they must have a *high frustration threshold,* so that repeated failures or confusing results can be faced calmly. Scientists think *strategically;* they develop optimal strategies for the direct straightforward solution of problems, but recognize that suboptimal strategies, using indirect approaches to problem-solving, are more realistic even if they provide only partial answers.

FOUNDATIONS OF SCIENTIFIC EXCELLENCE

Some of the internal elements of success in science as a *discipline*, as contrasted with science as an *occupation*, can be recognized in many of its practitioners. Although not uniformly represented in all careers, they often include

- *Depth of insights*
  - —Capturing those few luminescent moments when data coalesce to provide a clear image of something new
- *Conceptual thinking*
  - —Emerging above the cloud level of data, and
  - —Fitting a few pieces of a giant puzzle together in logical order and seeing a coherent part of the total picture
- *An innate urge to understand*
  - —Being dissatisfied with partial answers, but
  - —Recognizing and accepting the limits of available methodology, and
  - —Maintaining an "eagerness for inquiry" even when progress seems slow
- *Willpower and energy*
  - —Pursuing almost endless experiments and replications, and then
  - —Doing just one more experiment, one more replication, or taking a few more field observations, and
  - —Reading, digesting, and assimilating the welter of information in rapidly proliferating journals
- *Judgment*
  - —Examining a field of inquiry to determine where significant gaps in information exist
  - —Selecting research problems which are adequate in scope but still amenable to solution

—Asking appropriate questions, and

—Setting long-range as well as short-range objectives

• *Elegance of experimental design*

—Organizing research programs which have logic and beauty, and

—Being innovative when required by the research problem at hand

• *Perspective*

—-Seeing beyond the present experiments or observations, to those which may have further application in the future, and

—Comprehending where results fit into a larger landscape

• *Flexibility*

—Changing directions when enough facts so indicate

• *Absolute veracity*

—Retaining, in any professional activity, a sense of "obedience to the unenforceable"—a willingness to subscribe to procedures and actions voluntarily and not because of fear of the consequences of noncompliance

• *Receptiveness*

—Accepting and appreciating constructive criticism

• *A prepared mind*

—Evaluating and assimilating background information, and

—Expressing ideas and results superbly, both orally and in writing.

This list of internal elements of success is too short. Given a few moments' reflection, every professional could add to it or disagree with one or more of its components. It does, however, seem to contain elements represented repeatedly in case histories of successful scientists.

THE CRITICAL ROLE OF INSIGHTS

The preceding list of critical elements of success in science was deliberately headed by the difficult but essential term "insights." Insights are pivotal in the continuum of stages that we call scientific research. An idea or hypothesis leads to experimentation or other observations; findings are evaluated and additional experiments or observations are conducted; results of these studies are evaluated, along with findings of other scientists; and some synthesis is achieved (or additional problems are uncovered). At several points in this process (as well as subliminally throughout the process) an emergence of pattern or correlation may occur. The nature of emergence of such insights may vary. It may occur infrequently as a sudden lightning flash (the "eureka episode"); it may occur more frequently through the slow painful stepwise accretion of background data leading to the emergence of a final synthesis; or it may occur as the resolution of controversy, in which strong and variant opinions or hypotheses collapse into or are melded into a single concept. By whatever route, the final achievement is a higher level of understanding of natural processes—in some few instances a quantum leap from the existing level—and a closer approximation of that understanding to reality.

Science has been described as a creative process, consisting of external acts of data collection and analysis, accompanied by internal acts of hypothesis formulation, experimental design, introspection, synthesis, and concept development. In many ways this is similar to creative acts in the arts or other human efforts—a blending of the internal and the external.

The moments of insight which illuminate scientific existence are rare occasions of exceptional perception in a world of average events. These are times when investigators rightly feel that they are moving where the human mind may not have been before.

JOYS

No consideration of the internal journeys of scientists could be satisfactory without an attempt to convey, principally by selected examples, a little of the joy of "doing good science." Joy is a reasonable descriptor, but its connotations are too limited to express the fullness, the richness, the multiple pleasures that are found in the practice of good science. The descriptor has been overused for almost every pleasurable pursuit in present-day society, ranging from sex to cooking to smoking. Substitute terms such as "exhilaration" or "ecstasy" can help transmit impressions that science is a major sustaining and expanding force in the lives of most practitioners, and that there are "highs"—many of them—that often exceed those found in most occupations.

Science is a principal contributor to the belief that the human mind can transcend animal boundaries, to probe successfully in understanding natural phenomena. Thus, some of the "joys" are those of insight and discovery, of producing proof through experimentation and observation, of documentation through publication, of interactions with peers, and even of communication with the public. Among the joys are a sense of quiet satisfaction and well-being, and a feeling of internal security—knowing that it is possible to function effectively as a professional among professionals. Most of the real but often underestimated pleasures of doing good science are, therefore, internal ones, far removed from the prizes, awards, and promotions associated with public recognition.

Some of the joys of an internal journey are evoked spontaneously by the following:

- Completion of a long, exhausting, and good experiment
- The first faint trace of a new insight emerging from a previously confusing set of data

- The sudden and preferably unvoiced realization that you are an "authority"—that no one in the world knows more than you do about a particular subarea of science (however small)
- The day galley proofs arrive for your first published paper that you know is replete with good substantive science
- The first time you are introduced to a colleague who has actually read your paper and remembers your name
- The pleasure of really communicating a concept to a large class of undergraduates
- A standing ovation from the class on the last day of a large lecture course
- The good and well-prepared graduate student who deliberately chooses you as a mentor and thesis advisor
- Graduate students who, after earning their degrees with you, go on to do significant things in science—and remember to say thanks
- A superb qualifying exam or thesis defense by your current protégé
- Knowing that you are part of the mainstream when more than one of your papers are cited in the reference section of a colleague's paper
- Walking up the aisle of a session room after giving a paper that you know was great
- Chairing a panel session and knowing that you belong in that particular role
- Listening to a really excellent keynote speech given by a good friend and colleague
- Being selected by colleagues to head a professional society
- Hearing a truly exceptional oral presentation, after sitting through tedious hours of mediocre ones

- Participating in a review panel meeting to which you have clearly made significant contributions
- Submitting a research proposal that is accepted rapidly and enthusiastically, without nit-picking or recommendations for modification
- Completing a synthesis—assembling a great mound of data and seeing an ice sculpture emerge
- Published verification of your earlier research results by other competent professionals
- Reading a review paper which deals perceptively and positively with your published work.

Every good scientist would have, after only brief introspection, many additions to this list, but none would deny the reality of joy (or some appropriate substitute descriptor) as a principal component of the reward system of science.

Viewed a little more abstractly, some essential pleasures of science as a profession include those resulting from

- *Creation of knowledge that matters*—with its core the primacy of fundamental research
- *Recognition of the "elasticity of limits"*—of human ingenuity and available energy to carry out elaborate research, and of synthesizing ability to see patterns emerging from disparate data
- *Moving one step beyond*—in which a project that is very good in itself, and which would satisfy most criteria of excellence, is by one additional stroke of imagination, one further effort, one more insight, carried to the level of the outstanding by professionals with an inner drive for excellence
- *Being there*—participating actively in expanding the perimeter of knowledge, and in moving back the boundary of the unknown

In summation though, the pleasures—yes, the joys—of science can't be encompassed in any list. They are an integral part of the fabric; they transcend functional descriptions of science; they are realities to all the excellent scientists whose careers form the foundation of this book.

It seems appropriate to end this discussion of joys by giving brief recognition to the sorrows which are common even among those who might be judged most successful. Underneath the surface film of success there is almost always a hypodermis of sadness—for opportunities not taken or paths not explored. There can be persistent regrets for the children who grew up while laboratory activities took too much of each day, for the important senior mentors who slipped away and disappeared almost unnoticed in the rush to achieve, for the spouse who tried to understand but eventually abandoned the struggle, for the promising research leads not followed, for the frustrations of not getting well-deserved recognition, for the colleagues and assistants who were used and then dismissed as no longer relevant—all interwoven with the triumphs and rewards.

CONCLUSIONS

Science, then, is an internal journey of introspection, synthesis, and concept development, as well as an external one of experimentation, observation, publication, and recognition; it is an exercise of the mind as well as an interaction with the outside world. Even though science today is more and more a group effort, requiring the contributions of members of a research team, the best discoveries or insights are often those made alone, in silence or contemplation, after appropriate observations and analyses of available information have been made. Advances in understanding are usually incremental, and the increments are usually small but satisfying. The "great conceptual leaps" which titillate the news media are rare events, but even the possibility of such an event, on whatever scale, at some point in a professional career, adds to the pleasure of the journey.

REFERENCE

1. Abraham Pais, *The Science and the Life of Albert Einstein* (New York; Oxford University Press, 1982).

CHAPTER 6

DESTINATIONS FOR SCIENTISTS

The Basic Functions of Scientists—Research and Teaching; The Elitist Landscape beyond Research and Teaching; The Scientist as an Administrator, Bureaucrat, Politician, Entrepreneur, and World Traveler

INTRODUCTION

Examining the case history files for insights about those who have really made it in science, seven categories of professionals emerge (with some obvious degree of intergradation): (1) the research scientist, (2) the scientist-educator, (3) the scientist-administrator, (4) the scientist-bureaucrat, (5) the scientist-politician, (6) the scientist-entrepreneur, and (7) the international scientist. Categories (3) through (7) have been only partially identified and described in the published literature on the sociology of science—or in any other literature for that matter. They represent departures from the generally accepted role of the scientist as a research specialist and teacher, but all five emerge as career peaks for a surprising number of individuals who begin in the laboratory. Because of the frequency of occurrence of these latter categories as career destinations for scientists, and because of their weak public recognition, it seems important, after adequate attention is given to the research scientist and the scientist-educator, to consider each in some detail. A key point here is that individuals in categories (3) through (7) *usually still think of themselves as scientists.* Some still maintain a degree of scientific competence; many contribute—directly or indirectly—to progress in science; and all are part of the community of science. Another point is that many individuals in these latter categories have emerged from successful careers as active scientists, and they find success and pleasure in modified science-related functions.

We should begin with the stereotypes—the research scientist and the scientist-educator—still in the majority and the only categories consistently producing scientific data. We will then move quickly to the transformed scientist as an administrator, bureaucrat, politician, businessman, and world traveler.

THE RESEARCH SCIENTIST

The stereotypic scientist wears a white lab coat and is often pictured bending over a bubbling flask, peering at the readings on a complex analytical gadget, or talking to a computer. Some scientists still do those things with great intensity. To them, research is an all-absorbing state of mind coupled with a problem-solving outlook on life. Many research scientists are willing—even eager—to commit most of their waking hours, and even their subconscious mental processes during sleep, to the problem at hand and to steps toward its solution. This state of mind can occupy evenings and weekends; it can intrude upon and dominate almost any other activity; and it can on occasion become almost compulsive in intensity. *Preoc-*

cupation with problem-solving is clearly a principal distinguishing characteristic of the research scientist.

Coupled with this preoccupation is *dissatisfaction with the existing state of knowledge about natural processes and events.* Research literature abounds with admissions of inadequacies—in concepts and in data—for almost every scientific specialization. Such admissions are standard elements in introductions to scientific papers. While it is true that absence of information helps to justify doing the research (and thus gives the investigator a rationale for his or her existence), the ritual expression of incompleteness of knowledge reflects a basic attitude of research scientists. Dissatisfaction with the status quo must therefore rank importantly in any list of subspecies characteristics of research scientists.

Disrespect for and distrust of the obvious are also innate or acquired traits of the research scientist. A safe operating principle is that almost nothing is as it seems on first examination—a principle somewhat lacking in profundity, but one which is critical in its application to problem-solving in science.

Closely related to distrust of the superficial is a *belief in an underlying orderliness in natural processes,* if the investigator is bright enough and perceptive enough to detect it and describe it.

Coping successfully with variability in natural events must always be a major task for research scientists. Dealing perceptively with admittedly and inherently incomplete sampling is a perpetual problem, and the sampling problem leads quickly to the absolute necessity for and appreciation of *statistical procedures* in examination of *any* data set, even the simplest. Scientists must, without exception, either be statisticians or they must have intimate and continuing relationships with statisticians. Critical elements include acceptance of degrees of randomness of observed events, and assessment of the extent of replication necessary to reduce variability to acceptable levels.

Objective evaluation of the published research results and conclusions of others is another distinguishing trait of research scientists. Elements include necessary but often tedious repetition of critical experiments or observations of others, willingness to express disagreement with findings and conclusions of others, if warranted, and acknowledgment of contributions of others in concept and data-base development. Science is usually held to be "self-policing" in the sense that incorrect conclusions or spurious data will be apparent eventually through published results of colleagues. The reality is that, in an era of pressure to publish new and original findings, any repetition of the work or verification of the findings of others has relatively low priority and is often not done. The low probability of verification has led to instances of fraud in science—some of them reported with great glee and some consternation in the news media. Examples are discussed in Chapter 9 on "The Pathology of Science."

Conducting research and communicating findings to the scientific community are worthwhile satisfying goals for "hands-on" careers in science—unobscured by administrative and other responsibilities. Unfortunately, this ideal state is vulnerable to disturbances from the external environment. Research directions and emphases may be influenced by administrators; funding may be affected by committees or politicians; even the continued existence of some industrial research groups may be threatened by corporate decisions. Survival and well-being of research scientists often require their occasional participation in peripheral functions (administration, political interactions, public relations), usually at the expense of personal scientific productivity. The so-called "ivory tower mentality" characterized by freedom to do research of one's choice may exist, but it is a luxury available only to a few.

The word "pioneer" is overused and an anathema to most professionals, but Dr. Jeanne Lemieux, a federal scientist at a senior field grade, must be so labeled. She has over the past decade developed and applied an array of new techniques to approach a persistent and significant research problem related to the agency's objectives. She has done this almost single-handedly, except for a few technicians and temporary assistants.

Honors and credit have been slow to accrue to her but she has participated recently in several important national workshops, and two years ago she was invited to a symposium in Japan. In that same year she was named to a scientific committee of an intergovernmental research organization. She has published a number of papers—some of them in rather obscure journals but all of them very well written.

The laboratory to which she belongs has existed for 30 years, and has developed extensive information of importance to industrial development. The rationale for continuation of the program has been questioned recently, as part of a program of reductions in some federal agencies. Dr. Lemieux naturally has serious apprehensions, and some concern about the assessment of her scientific contributions, when decisions which are principally political can determine the continued existence of research in that area.

Her response has been to reassert the worth of her contributions—to the agency and to the general public. She has accepted the requirements of the agency to supply planning documents and interim reports; she has developed some rapport with key members of the industry group most affected by her research; and she has even discussed the significance of the research with legislative assistants who may contrib-

ute to funding decisions. Her professional metamorphosis to meet a perceived threat has reduced, but has not eliminated, scientific productivity.

It is apparent from this and other examples drawn from case history files that the research scientist, totally committed to his or her work and unimpeded by extraneous forces, exists, but does so principally as an ideal and rarely as an actuality. The pure strain is tolerated only in extremely limited numbers, somewhat like highly inbred strains of laboratory mice, which exist for the specialized but narrow purposes of the experimentalist. The research scientist is treated in some ways similarly to the artist and the craftsman of previous centuries—as a luxury supported by a society able to invest temporarily in his or her specialized competence.

Dr. Nancy Calvert is a member of the staff of a small, highly specialized federal laboratory in a delightful rural location. She has been conducting unassisted (at her preference) laboratory-oriented research in her specialty for 15 years, and is recognized nationally and internationally as an authority. She has published two books and over 50 scientific papers; one of her books will undoubtedly be a classic of its genre.

During her government career she has deliberately refused any kind of administrative responsibilities, except for the required minimum of planning and progress reports. Despite this, she has progressed to a salary level equivalent to that of the assistant laboratory director—demonstrating that it is possible within the federal research structure to achieve financial recognition based on merit and sustained productivity.

The scientific community beyond the laboratory has endorsed her competence and productivity. Two years ago she was elected president of the major international scien-

tific society in her specialty area, and she has been an invited speaker at several international symposia. She has been a vigorous proponent of equal rights for women in science, and has participated actively in regional EEO programs.

Dr. Calvert has made and will undoubtedly continue to make significant contributions to knowledge in her field. She has been encouraged and protected by scientist-administrators who know that excellent research must have high priority, regardless of other agency pressures.

Beyond the government scientists described so far, a special category of investigator, worthy of respect and attention here, is the *university research associate*. The nonteaching professional component of the university staff treads a careful path between prosperity and extinction. Some of the larger universities offer benefits and salaries to research staff members that are commensurate with those available to the teaching faculty, but the standard practice requires research staff members to scrounge for most or all of their salaries through grants. This stress-provoking situation is tolerable to the few; the many soon depart to government or industrial laboratories, or accept a teaching position with some job security at a lesser institution.

We have witnessed in recent decades the emergence of research institutes, with direct or indirect academic affiliations. These institutes, often in rural or coastal locations, encourage full-time research, with minimal student contact, usually through short courses or brief lecture series in ongoing courses, or the occasional graduate student. The dark side of such institutes is their dependence on public or private funding sources—a dependence that is transferred quickly to the research staff. Survival and well-being thus become contingent on skill in grant proposal preparation and in selection of research projects of interest to funding agencies.

Dr. Ambrose Licht decided early in his career to concentrate his research efforts in a narrow but fundamental aspect of science. His entire career of almost 30 years could be considered an expansion of his Ph.D. thesis topic; he is now widely considered the world authority on that circumscribed subject. After receiving his degree he moved to a large research institute of a major eastern university, where he has remained as a research associate, producing substantial contributions to the literature, reviewing manuscripts, contributing chapters to books, and increasing world knowledge in his field of specialization. His student contacts, by his own choice, have been minimal, although he serves frequently on thesis committees and advises an occasional highly selected and totally dedicated graduate student. His involvements in academic committees and scientific society affairs have been correspondingly minimal—again by his choice.

Fortunately, the funding for the institute has been stable, with a substantial core from private sources. Fortunately, also, the institute is one of the few with tenure for outstanding research professors, so his position is remarkably secure in an insecure scientific era.

Dr. Sidney Pollack represents an emerging group of academic scientists employed by institutes formed by university consortia. His primary responsibility is research, but he has faculty affiliation with several member universities, and offers occasional graduate-level courses in his specialty. Like many institutes of its kind, his is located in a pleasant rural setting, with excellent grant-purchased equipment. Staff members form a critical scientific mass, with remarkable intragroup cohesion and compatible interests.

Dr. Pollack is a compendium of good traits—intelligent and energetic, a hard charger and early achiever—and he produces a continuously high volume of research papers, usually in collaboration with associates, assistants, and graduate students. He is active in regional scientific affairs, preferring to place emphasis here rather than in national activities (although he has participated effectively in several international scientist exchange programs).

One of the darker aspects of Dr. Pollack's career is the uncertainty of the future. As is often the case, the consortium provides minimal funding for the institute, which requires him to invest a substantial part of his working day in a search for additional grants and contracts—which in turn requires that much of his research be tailored to meet interests and stipulations of granting and contracting agencies. The job future, even for a proven producer of his caliber, is questionable, since staff members are outside any member university's tenure system. Despite this cloud, his present consists of exhilarating simultaneous participation in several research projects, classroom and laboratory teaching, and an outstanding record of paper production—interspersed with occasional short-term research fellowships and exchanges in foreign countries.

Dr. Pollack represents a modern phenomenon in research—an exceptional scientist with good credentials in a system deficient in security. His security at present is internal—in knowing that he is very good. It seems probable that he will soon move to a larger, more stable research environment, even though the nonprofessional advantages of his present location are strong incentives to stay.

THE SCIENTIST-EDUCATOR

The principal strike force of academic science consists of scientist-educators at various grade levels from instructors to professors, who perform daily feats of magic and endurance, carrying on original research interspersed with lectures, laboratory sessions, student conferences, seminars, faculty committee meetings, trips to the supermarket, and mowing the lawn at home. Most do all these things remarkably well, and with enthusiasm. Some receive appropriate recognition and rewards; some do not.

We hear so much lately about the down side of academic existence—the shrinking student body, "nontenure track positions," the scarcity of grant funds—that it is a relief to examine case histories of those who have been successful as scientist-educators in this difficult current environment. True, the roots of some of the success stories extend back to a gentler era when many of the current stringencies did not exist, but they should still provide insights about aspects of the job that are important to success.

Professor Allen Queensbury exemplifies the academic who does several things well and with enthusiasm. A graduate of a small undistinguished college, with a Ph.D. from one of the best universities in the country, he was selected from countless candidates for an entry-level position at an Ivy League school. His course offerings were substantive, if not inspiring, and his original research was adequate and extensive, but his particular interests and skills were in digesting and presenting in book form the accumulated knowledge in his area of specialization. He produced, early in his career, outstanding books for professionals as well as interested amateurs—books which were and are authoritative and informative, as they have appeared in several revisions.

Professor Queensbury developed, simultaneously, an interest in summer field programs for undergraduates—programs which required combined expertise from a number of universities. The programs grew and coalesced into a mul-

tidisciplinary summer institute, which is today recognized as one of the best and most stimulating of its kind. He is still its principal motivating force, but others have joined and have assumed much of the responsibility for continuity.

His third principal vocational preoccupation (in addition to books and the summer institute) has been planning and organizing workshops and symposia, to review, evaluate, and improve the status of knowledge in his specialty. Several of these efforts have resulted in a continuing series of edited volumes, and few symposia are complete without his contribution.

Related academic affairs have benefited from his reasoned participation. He heads a select faculty committee charged with keeping under constant surveillance policies affecting the undergraduate curriculum, and was recently chosen to head a faculty review committee on ethical procedures in research. Truly a man of many parts, Professor Queensbury clearly typifies the successful scientist–educator sequence.

As part of casual background research for this chapter, 20 scientist-educators, considered by colleagues and students to be good teachers, were asked to jot down a list of 10 criteria by which they would recognize excellent scientist-educators. The priorities varied, but some of the more consistent criteria were these:

- They are enthusiastic about science, and convey to students the excitement and satisfactions of scientific research.
- They consider the factual content of a course to be basic and necessary, but feel that they must present it with stress on the transiency and incompleteness of currently accepted views and beliefs.

- They search for and nourish creativity in students, through guidance in independent studies and one-on-one discussions.

Dr. Daniel Reilly is my all-time favorite for sparking and maintaining a spirit of inquiry among students. He has been a faculty member at a large eastern university for 18 years, beginning as an assistant professor when the department consisted of four members. He has tenure and has served a three-year rotational term as department chairperson. He teaches by choice a large introductory undergraduate course, a senior-level course in his specialty, and a graduate seminar course.

Many external signs of success are apparent in his career. His rate of promotion has been better than average, and he has achieved recognition from colleagues within the department and in the larger scientific community. He is one of a significant percentage of faculty people who elect and can carry off close personal relationships with undergraduates as well as graduate students. His great enthusiasm for his specialty is transmitted to students; his genuine concern for student growth in science is evident; and his frank, non-condescending, but demanding classroom and laboratory manner produces results.

Professor Reilly was asked recently if he wanted a second three-year appointment as department chairperson. He declined in favor of having more time for research and graduate student contact. He has received research grants from public and private funding sources; his grants support a secretary, two assistants, two postdoctoral fellows, and two technicians. He invests about 10 percent of his work week in preparing new grant proposals, and 5 percent of his time in reviewing others' proposals and in participating in site review teams. Last year he was invited to present a series of lectures at the Royal Society of London.

Additional criteria for recognition of excellent scientist-educators are

- They expect to serve as role models for students, and take the responsibility seriously.
- They make the point repeatedly that the evolution of science depends on many small contributions, which lead eventually to unifying concepts, great and small.
- They develop and maintain rapport and informal communication with students at all levels.

Father figure, friend, advisor, provider of $10 loans in crises, confessor—all these descriptors apply to Professor Ken Lee who is also an exemplar of a small proportion of university faculty members who have taken a Ph.D., then stayed and prospered at the same institution for their entire academic careers. Some universities scrupulously avoid hiring their own Ph.D.'s, considering it "inbreeding," but others (like Dr. Lee's) seem to have no strictures of this kind. The factors leading to remaining, in many instances, may have been entirely fortuitous—the resignation of a junior faculty member with the same specialty, the retirement of a major professor with that specialty, the expansion of a department (rare today), or the evolution of a short-term postdoctoral appointment into a more permanent instructorship.

Professor Lee was one of the fortunate ones—receiving his Ph.D. in the year that his thesis advisor announced his resignation. Being exceptional academically and as a teaching fellow, he competed successfully with a horde of applicants from outside for the position of instructor the following year.

His movement through the musical chairs of faculty grade levels was not meteoric, but it was steady, enhanced by effective undergraduate teaching. Moreover, excellent rapport with undergraduate and graduate students, preparation and guidance of good Ph.D. candidates, substantial personal research, and enthusiastic participation in selected faculty committee activities, all contributed to his success. His natural gregariousness and enthusiasm led in part to election to offices in a carefully limited number of scientific societies. This visibility, plus his own scientific accomplishments, in turn led to leadership of several successful national symposia and workshops. His faculty committee activities led to close mutually supportive relationships with members of other departments and schools within the university.

Still other criteria for recognition of excellent scientist-educators are

- They consider excellence to be achievable, and expect it at any level of instruction.
- They consider students to be equals and partners in the transfer of information.
- They are dynamic lecturers who work at maintaining student interest and involvement.
- They have time or make time outside class hours for private informal discussions.
- They are secure about their knowledge of the subject matter, and are relaxed and forthcoming in discussions of it with others.
- They emphasize concepts but insist on a solid factual foundation for all science students.
- They encourage constructive criticism.

- They maintain as effective a balance as possible between instruction and personal research.

The list of proposed criteria for recognition of excellent scientist-educators (as developed by respondents to the survey request) went on, but some of the responses became mildly repetitive and only variations on one or another of the items considered so far. Some of the less cooperative of the respondents wanted to turn the exercise around and to list characteristics of *poor* scientist-educators (lack of enthusiasm for science, rigid and dogmatic, not current in knowledge of the subject, totally immersed in research, no time for students outside class hours, often away at meetings and conferences, talked only to God and personal favorites among the graduate students, reluctant to explore concepts or hypotheses, sarcastic and not helpful, superior and condescending attitude, etc.). The list of negative criteria might have been more effective in making the point, but, as I tried to explain to the rebels, the search *is* for excellence, and not for failure.

Investigation of the excellent scientist-educator should not be truncated or minimized. Critical issues should be addressed here—such as the conflicting demands of teaching and research (each of which could be a full-time job); the appropriate extent of faculty committee participation; the reward system for fundamental contributions to knowledge; and the financial pressures for off-campus consulting activities. Balanced commitments to the development of graduate students, to the larger scientific community as represented by professional society meetings and workshops, to extensive field studies, to reviewing and editing scientific papers of others—all require attention and discussion, but to explore each of them in depth is beyond the scope of this book.

What we *can* do is choose among the many issues to be explored, selecting a few for brief probes. One that has high priority for a number of scientist-educators is *achieving the proper balance between teaching and research*. Excellent teachers must make a maximum commitment to transfer knowledge effectively. They do this with the many methods mentioned earlier—enthusiasm, personal contacts with students, availability to students, maintaining competence, providing a role model—all of which reduce the time and energy available for contributions to expansion of knowledge through personal involvement in research. Some successful scientist-educators have found at least partial solutions to what seems like a dilemma. Examining available case histories, some obvious and some less-obvious approaches emerge:

- A research area can be selected that has sufficient breadth so that graduate students can be offered pieces of it for thesis topics. (The scientist-educator, however, keeps the core intact so that his or her personal publication will not be delayed awaiting completion of a thesis.)
- Grant acquisition should have high priority, so that research associates, assistants, or technicians can be hired. This implies entry into the fascinating, frustrating world of grant-chasing, but there is no easy alternative. The nucleus of a grant-supported research group must be paid employees; students are important adjuncts, but they are transients and must remain as adjuncts. A key is the employment of a good research associate or senior laboratory technician (or both, with luck).
- Graduate students can be employed in grant-supported research, provided there is a clear understanding about what part of the research (if any) is to be used in their thesis, and what part is to be used in publications by the professor. Multiple benefits accrue on both sides; the professor gets his research done, and the student

becomes part of a research team—an apprentice for larger postdoctoral activities.

- Undergraduates can also be employed in grant research, and such employment should provide motivation and insights about careers in science. It is important, though, to outline the terms of employment clearly, so that students understand their role, and do not have mistaken ideas about "instant professionalism." The merits of undergraduate involvement in grant-supported research are many, but the commitment of the scientist-educator is also high, especially in conveying attitudes, necessity for procedural precision, statistical treatment of data, and proper care of experimental animals. Undergraduates can be supervised by paid assistants or technicians; but their responsibilities must be defined carefully.

- Research which involves field operations can be especially effective as a teaching device as well as a producer of scientific data. Many field programs include extensive repetitive observations which are well within the capability of undergraduate assistants. A superior field research team can engender a group spirit and a sensitivity for incremental contributions to knowledge far beyond that normally found in the laboratory. A common thread in many case histories of scientists is the importance of participation in just such a field program to a decision in favor of a career in science.

Another issue, closely related to the previous one of achieving a proper research/teaching balance, facing the scientist-educator and chosen arbitrarily from the long list waiting for proper discussion, concerns *the reward system for contributions to knowledge.* The institutional rewards are tenure, elevation in rank, and more pay; the professional rewards include status and credibility as a scientist, prizes and awards, and society offices. The institutional rewards are most tangible, and are based theoretically on achieving a proper (from the institution's perspective) balance of research and teaching. The location of the fulcrum for this precarious balance is of course largely determined by the institution. Most major universities expect and insist on significant continuous contributions to knowledge from their faculties. Smaller public and private colleges often place original publication at much lower priority than teaching (giving some credit for production of textbooks which evolve from course offerings of individual faculty members).

Even among institutions valuing original contributions to the literature, there is continuing disagreement over judgments of the worth of those contributions. How, for example, in a decision about promotion to an associate professorship, can the published results of 10 years of thorough exhaustive investigation of mud snail ecology be compared with a brief but brilliant theoretical paper on predator–prey relationships in the sea? How, for another example, does the editing of a book of contributed papers rate when compared with a substantive original paper? How does publication of a critical review of present understanding in a subdiscipline compare with that same substantive original paper? Answers are not simple, except that scientists should be involved in the decisions as well as administrators. Fortunately, many tenure review committees do include such representation.

To retain some balance in this chapter on destinations for scientists, it is time now to move on to other hyphenated categories. Before leaving the scientist-educator, though, it seems appropriate to point out the pivotal nature of his or her job— stimulating, inspiring, guiding the best of the new crop, while contributing actively to the fund of information to be transmitted. Most successful people in this category are convinced that they have the best of two worlds—teaching and research.

THE SCIENTIST-ADMINISTRATOR

Many successful scientists evolve gradually or rapidly into administrators—of university research groups, of government laboratories, of granting agencies, or of industrial research divisions. Some go willingly, eager for the prestige, power, and salary that are concomitants of the job; others go a little reluctantly, occasionally unknowingly at first, often with the feeling or conviction that research can be best accomplished if they "sacrifice" some of their time and careers to scientific administrative functions.

Prodigious research productivity, exploiting a few significant insights, proved to be an effective route upward for Professor Joan Cranwell. Working in a specialty where technician support is essential to productivity, she began with energy and competence to explore in great depth several research areas where others had only paused fleetingly. With a small initial core of assistants, and with a continuing sequence of exceptional graduate students, her annual production of substantive research papers sometimes exceeded 20 per year. Not at all laboratory-bound, and being politically astute (in the broad sense), she participated extensively in scientific society activities—with a particular bent toward organizing and chairing workshops, panel discussions, and symposia. Research contracts and grants came in increasing sizes from several agencies, and since part of her research required major computer resources, she moved an entire support group and her own activities from one university with minimal capabilities to one with excellent facilities and an interest in expansion. The move was mutually advantageous; the recipient university gained prestige and reputation, and Professor Cranwell gained the administrative approval and playing room for her ever-expanding research interests. Unfortunately (or fortunately), the new environment also provided scope for expression of her managerial, organizational, and political talents. She became coordinator for several interdepartmental research programs, and was offered the directorship of a substantial university institute—a wholly administrative position. Interestingly, it was at this point that her overriding interest in close association with active research modified what might have otherwise been a predictable transition to full-time administration. She withdrew her name from consideration as director of the institute, even though she was clearly the most obvious choice. The full story is still in the future. Her broad research interests, her multiple research-associated activities, and her scientific paper production, continue unabated.

Good administrators with scientific credibility are clearly needed for today's complex science establishment. Research has not been for a long time an exclusive preserve for individuals functioning independently (although some still persist). Organizations exist to direct and manage research, and their effectiveness and productivity depend in part on the skill with which they are administered (recognizing that good scientists performing effectively within those organizations will always be the ultimate determinants of success).

Director of a major field research group of over 100 people, Dr. John B. Perelli typifies the successful research manager who has achieved much, but who still has substantial growth potential. Still in his late forties, he is recognized as an expert in his discipline, and is a frequent participant in national and international forums. He was recently elected chairman of a prestigious international committee and was appointed to several others. Although management-related activities take up much of his normal day, he is able, by extending his working hours to include evenings and week-ends, to maintain professional competence, to write scientific papers, and to interact superbly with an extensive world-wide network of colleagues and friends.

He does the national and international science routines at some cost, however, to his day-to-day contact with the research staff, who rarely find him available for the leisurely discussions which are so important to development of cohesive research groups and to the evolution of intermediate-level supervisors. A partial solution has been effective delegation of those responsibilities to an excellent group of project managers, who have assumed the day-to-day hands-on supervision.

In goverment research, assessment of aspects of work performance considered critical to promotion or job retention must be made continuously. To a large extent, this assessment must take into account the wishes of an immediate supervisor, regardless of the level of the position. Dr. Perelli has read, correctly, his supervisor's interest in and concern for involvement of his key staff members in international science, and his supervisor's expectation of exceptional performance in the difficult dual role of research authority and effective manager. Dr. Perelli has the intellectual ability, background, and energy to do all these things well.

Dr. Perelli has become, almost intuitively, a master at "networking." His voluminous correspondence and massive telephone communications (always with substantive matters to discuss) have helped to create national and international visibility among peers and colleagues for the good work which he directs. These communications have brought him recognition in the form of committee, workshop, and symposium assignments. He typifies what might well be termed a "world-class scientist."

Scientist-administrators can be considered as part of a select group—along with scientist-bureaucrats and scientist-politicians (to be discussed later)—essential to, and not merely an adjunct to, modern science. If this premise is accepted, then it would be useful to identify some basic tenets of the trade, some operating principles acceptable to members of that group, which lead to success. Groping through the impressive amount of individual variability which exists, it is possible through extensive culling to find a principle here and there, among which are

- Scientist-administrators should have credibility as scientists. Professionals within a research organization do not want funding and management decisions that affect them to be in the hands of accountants or ribbon clerks.

Dr. William G. Harris is an excellent example of the successful scientist-administrator. He is the long-term director of a major state-supported but partially autonomous research laboratory, with a staff of 200—over half of whom are professionals, and one-quarter with a Ph.D. The laboratory is a teaching and research institution, with university affiliations, with about 100 graduate students, and with staff competence in many subdisciplines.

Dr. Harris is an excellent scientist, who enjoyed credibility and an international reputation in his specialty before his appointment as director. He has maintained his scientific involvement, principally through graduate students and grant-supported research assistants. Joining the laboratory as a research assistant professor, he, subsequent to his appointment to the directorship, has systematically, over 20 years, built the group from a 25-person single-discipline organization to a 200-person multidisciplinary research and teaching institute. He has done this by (1) judicious selection of staff, (2) persuasion of the state legislature to make major funding commitments, and (3) acquisition of federal and other research grants and contacts.

Dr. Harris has "charisma"—an excellent descriptive term despite its recent overuse. He is physically attractive, has a forceful personality, is very intelligent, and is an excellent speaker. His instincts for correct actions are remarkably good.

His negative traits are pale by comparison. Some staff members find him autocratic and dogmatic. He exhibits some reluctance to delegate authority to associate and assistant directors, and he has an obvious impatience, sometimes roughly expressed, for mediocrity and nonproductivity. His enthusiasm for expansion has at times led to overextension of the funding base of the organization, with associated trauma.

Dr. Harris has some characteristics which fit him also for the appellation "scientist-politician." He is willing and able to talk with legislators and other elected officials, to keep them informed, and to present information and requests for support in a very effective way. Furthermore, he has characteristics that politicians appreciate—the ability to sort out meaningful information from trivia, and to take decisive action based on the best available data.

Other operating principles for scientist-administrators include these:

- Scientist-administrators must acquire professional status in management/executive areas. Competence in science is only one background requisite for good management of research organizations. Expertise in personnel management, organizational planning, delegation, finance, psychology, and public relations is of equal stature, and is expected by professionals within the organization. Acquiring such expertise usually occurs at the expense of continued active participation in research and teaching.
- Scientist-administrators should make some attempt to maintain contact with one or more discipline areas, but should not expect to be active participants in hands-on research. Some degree of competency can be retained by reading, by discussions with active colleagues, by attending occasional professional society meetings, or by preparation of infrequent but good overview papers.
- Scientist-administrators should always keep in mind that the success of the group depends ultimately on the conduct and publication of good science. Those actions which encourage such ends should be the ones which receive greatest attention. Effective management is needed, but the group exists to produce relevant information, and not merely to be used as a model for good administrative practices.
- Scientist-administrators must remain concerned with the *substance* as well as the *form* of science if they are to stay alive professionally. Too often the form becomes the sole occupation—in the sense that planning, organizational politics, fund-raising, and public relations occupy an inordinate amount of attention.

The scientist-administrator blends into the next category, the scientist-bureaucrat; the distinction between administrator and bureaucrat is a fuzzy one with a large degree of overlap. Some subjective criteria might include these:

- Administrators deal principally with people; bureaucrats deal principally with documents.
- Administrators sign documents; bureaucrats prepare them.
- Administrators occupy line jobs; bureaucrats act as staff.
- Administrators plan and act; bureaucrats discuss.
- Administrators organize and lead committees or conferences; bureaucrats participate in them.

THE SCIENTIST-BUREAUCRAT

The population of federal bureaucrats working in the District of Columbia was 366,000 individuals by recent count! This is more than the total number of scientists in the entire United States (313,000) according to a recent count using criteria established by the National Academy of Sciences (quoted in Harriet Zuckerman, *Scientific Elite: Nobel Laureates in the United States*[1]). If government bureaucrats working outside the District, and bureaucrats on state and local levels are added, and if their numbers are augmented by academic and industrial bureaucrats, then the disparity becomes overwhelming. Some of those bureaucrats have science backgrounds, and some function effectively and happily in the bureaucracies of science.

Few subjects in the sociology of science are more fascinating than the scientists who have metamorphosed into bureaucrats. In reviewing the many scientific case histories which form the statistical base of this book, it became apparent rather quickly that *some of those people who call themselves scientists should properly be classified as "bureaucrats"*—in the broader definition of the term. With this perception, it is reasonable to ask if scientists bring unique contributions to their bureaucratic roles, as compared, for example, with lawyers or accountants. It is also relevant to find out if scientists have greater survival capabilities within the bureaucratic subculture, and if they can undergo such a metamorphosis and still carry on any functions which could be classified as "science." It might even be interesting to see if the metamorphosis is reversible, in at least some instances. Insights into all these problem areas lurk in the data at hand.

A common attitude, expressed repeatedly by scientist-bureaucrats, is an odd mixture of pleasure with the new challenges at the organizational fringes of science, occasional longing for the simpler existence of the laboratory and/or classroom, and fleeting well-camouflaged sadness over the decay of professional competence.

Scientist-bureaucrats must of course be immediately segregated into "career bureaucrats," dependent on ability, productivity, and seniority for survival, and "politician-bureaucrats," dependent on political appointment and political support for survival. Career bureaucrats, whether possessing a scientific background or not, provide "institutional homeostasis" so critical to any organization. Politician-bureaucrats, whether scientists or not, are transients, capable of extensive but short-lived effects on the organization, and programmed to self-destruct with a change in administration.

A Condensed Survival Kit for Scientist-Bureaucrats

Since this is a book advertising the "joy" of science, it is perfectly reasonable to inquire into the career satisfactions of scientist-bureaucrats. Obvious and superficial responses to the inquiry would be "survival and transient power" or "the challenge of manipulation of a complex system" or "interactions with exceptional bureaucrat and politician colleagues."

Achieving the satisfactions of survival, power, and well-being depends on awareness of multiple nuances of organizational behavior, some of which are unique to bureaucracies and any of which can be critical to a career. Listed here are some of the more important considerations, of which seasoned bureaucrats are aware:

- *The overriding importance of "paper."* To bureaucracies, words constitute actions, and documents are ends instead of means. Paper is thus the life-support system of any bureaucracy. The reality of this dictum is most apparent in government bureaucracies at the time of a change in political administration. This period of feverish activity is characterized by widespread re-creation and proliferation of documents. Everything from the previous administration, good or bad, must be discarded. New initiatives with catchy acronyms must be developed; new planning documents must be created immediately by intense staff activity; new goals and objectives must be enunciated which follow the dictates of the administration now in power.

  This delicious period of reorganization, rethinking, and restructuring has remarkable parallels with the periodic rewriting of history practiced in the Soviet Union. Policies and programs of a previous administration disappear; planning documents evaporate or are dismissed as ineffective or misconceived; new advisory committees are created, with new guidelines; files of the "old crowd" are relegated to remote storage buildings—in brief, the bureaucratic world is restructured to perform as if everything done in a previous administration does not exist and never did.

Good career bureaucrats—survivors—know about and expect such changes. They accept the reality that agency documents which were filed next to the Bible yesterday are today shredded, never to be resurrected or even discussed, except in derogatory terms. Meanwhile, the entire staff is happily engaged in creating new and equally meaningful pieces of paper, which will be discussed and modified, and will serve as operational dicta until the next change in administration.

- *The insertion of a new unit leader in bureaucracies.* Other aspects of organizational behavior are equivalent in importance to the veneration of paper. One of the most fascinating (and one most critical to survival) is the redistribution of power which follows the appointment of a new leader at upper hierarchical levels. The process can be simple, direct, and brutal, or it can be intricate, delicate, and prolonged. A few of the more common approaches are

    1. *Direct and often abrupt appointment of a new leader to an existing unit,* by political appointment or by career selection with "political sensitivity" (which means that the appointee's name is floated past appropriate legislators and other categories of politicians, with final approval hinging on their concurrence). The previous administrative head may be reassigned to another unit, placed in charge of a smaller field group, given the option of early retirement, or kept in a staff role with zero authority.
    2. *Major agency reorganization* is a favorite when a new administration assumes control. Almost all the executive-level people who were appointed by a previous administration are vulnerable—

especially if of the opposite party, or appointed by politician-bureaucrats of the opposite party.

Internal reorganizations can affect part or all of the agency. They provide opportunities for recombinations of units under new heads, erosion or elimination of power of certain existing heads, and dispersal or elimination of previous power figures by exile to field offices, or relegation to staff duties.

3. *A stepwise reorganization,* in which the power of one or more individual administrators may be enhanced at the expense of others. In *step one* a small unit, consisting of several pieces of preexisting organizational units, is designated, with a new administrative head and with a loosely defined mission. The new unit usually reports to an executive well up in the hierarchy of the organization. In *step two* the small unit absorbs several other preexisting units, assumes operational responsibilities, and soon is no longer a small unit. Heads of the absorbed units retire, disappear into other organizations, or continue for a time in the new group with much-reduced authority. The mission of the unit is defined and expanded. In *step three* the unit (now a major operating entity) absorbs the remainder of the preexisting organizational entities, with the departure or reduction in power of the administrative heads of those entities. The new administrative head now controls all or most of the components of the original organization, even though they may be variously restructured.

It should be noted, though, that reorganization may be for positive reasons too, particularly to provide additional upper hierarchical niches to permit promotion of good people. This reorganization device is especially effective in agencies with long-term well-established tables of organization. (It might be added that reorganization can be proposed as a panacea for almost any bureaucratic ill—low productivity, poor morale, high turnover, poor communications, etc.)

- *Elimination of an incumbent.* Removal from a bureaucratic position can occur for reasons unrelated to productivity or accomplishment—for purely personal whims of those at higher supervisory levels or as part of a larger reorganization which eliminates certain jobs. Perceptive bureaucrats recognize the realities of the system and develop unobtrusive but effective ways of coping with crises. They turn to friendly politician-bureaucrats or politicians, or to sympathetic industry representatives for advice and assistance, and they perform the remarkable contortion of keeping rear ends low, ears to the ground, and eyes on events.

  Bureaucrats also recognize the many devices available to expedite disappearances—and are acutely sensitive to any attempted application of these devices to their own situations. Some favorites are

  —Reorganizing the group and eliminating the functions of a particular position
  —Transferring positions or programs to less-desirable locations (Kodiak, Alaska; Pierre, South Dakota; Guam)
  —Moving a person into a new job classification, then after a suitable waiting period, eliminating that category

—Reorganizing the group and having the person report to a relatively junior supervisor

—Creating a new program, then claiming budget problems and eliminating it

—Removing the line functions of a position, and then making its staff functions impossible to fulfill

—Looking for and documenting the tiny illegal acts of which most people are guilty (favorite examples: tardiness, private use of official phones or copying machines, expense account violations, etc.)

• *Coping with the outside world.* Proper responses when the wolf pack closes in—when industry lawyers, concerned citizens groups, or politicians sense a minor flaw in the bureaucratic armor and smell blood—include some or all of the following:

—Public hearings at which the best and the most articulate scientific talent is arrayed to defend the establishment, to present reasoned factual data free from hesitancy or fumbling.

—Willingness to admit minor errors, as long as the integrity of the system is preserved—but a concomitant willingness in severe tests of the system, to sacrifice clearly culpable colleagues. [There is a built-in risk here, though—in that a senior bureaucrat must depend on a cooperative infrastructure. Part of any loyalty which may exist in such a group hinges on the expectation that those above in the hierarchy will protect (within reasonable bounds) those below.]

—Another less effective but widely used response is "stonewalling"—preventing or delaying access to information or opinions by the many bureaucratic circumventions which exist, such as requiring extensive approvals within the system before release of data.

- *Preserving core programs.* Budget fluctuations are of overriding concern to bureaucrats and bureaucracies. In times of budgetary stress (an almost annual event in today's climate) barricades must be thrown up around core and critical programs—leaving all other programs outside the barricades and therefore vulnerable to dismemberment or death. During those stressful episodes, some programs may be hidden by reorganization, even if program recombinations produce strange partners.
- *Implementing a new program.* Bureaucracies occasionally require rejuvenation—either by reorganization or by creation of new programs. Unfortunately, the "organizational homeostasis" discussed earlier has a big brother which can be called "organizational inertia" or even "organizational resistance." It is difficult to introduce new programs, especially when surrounded by budget restrictions, yet new programs can fill obvious needs or revitalize an existing group.

Scientist-bureaucrats spend significant chunks of their time planning, developing, and implementing new programs. They recognize that even the best proposals require careful preliminary sales efforts, or they will be filed and forgotten. They recognize too that effective program development can be an important ingredient in their performance evaluations.

New programs obviously require new funding. New funding obviously requires support from the administration in power and from the legislature. Some pathways include

—Active support from representatives of the party in power
—Assistance from aides of key legislators
—Endorsement by key agency officials

—Rapid achievement of critical mass in programs and people

- *Drawing a line.* Bureaucrats often carry the stereotype of compromise and inability to take firm stands. This is generally a true assessment but is a little unfortunate, because exceptional bureaucrats are on occasion willing to say "no" or "yes" or "this is crap" without equivocation. Such behavior cannot become a way of life, of course, but hard decisions and firm statements need not be totally foreign to bureaucratic existence. Line-drawing actions can be taken in a civil ego-preserving manner; occasionally such actions occur. The research on which this book is based has disclosed a number of instances among the best scientist-bureaucrats where principles have been preserved, despite perceived threats of negative career consequences.

THE SCIENTIST-POLITICIAN

Beyond the perception of "politician" as denoting "a participant, by appointment or election, in the public administrative power structure at local, state, or national levels," there is a broader connotation of the term which applies to many successful scientists. Scientist-politicians do far more than interact with the "political system" in the narrow sense. They are strategists, who plan and analyze any significant activity. They are public relations conscious, examining the effects of their actions on colleagues and on that part of the population which might be interested or affected. They are manipulative, taking full advantage of external opinions and opportunities to shape future activities. They are gregarious, seeking the company and companionship of colleagues and others in positions of

power or accomplishment, regardless of their fields of expertise. They are self-assured, acting like professionals and
expecting to be treated as such. They have high internal feelings of self worth and accomplishment, and project these feelings. They have excellent analytic and synthetic abilities,
which often transcend their own area of expertise.

Scientist-politicians seek out and enjoy activities on the
public perimeters of science, and often find particular pleasure
in acting as spokespeople for their less vocal but equally capable colleagues. They emerge in all kinds of public forums—
from early morning talk shows to late evening mass meetings
on environmental or other issues, and from roles as advisors
to small-town mayors to participants in expert panels of the
National Academy of Sciences.

---

*Some academic deans are noted for an ability to keep their
feet firmly planted in academic administration and in science. Others metamorphose completely and occasionally
superbly into administrators or politicians. Dean John Knorr
is an example of the former; from a base of productive
research and broad integrative writing in his specialty, he
moved from a distinguished professorship to a newly created
post as dean of a new school in a university of modest size.*

*He has demonstrated several critical abilities not com-
mon to most academics but important to success in his present administrative role. He is astute in financial matters and
is an effective fund-raiser; he can mediate internal disputes
with great skill; and he can inspire the mediocre to greater
exertions and the superior to produce outstanding
contributions.*

*His real expertise, though, is in political interactions.
He deals effectively with state political figures and state
bureaucracies, and has developed rapport with Congressional incumbents and staffs, as well as with the many*

Congressional subcommittees whose actions affect the fund-
ing and direction of science.

    Interestingly, he is still listed among the broad think-
ers—the "conceptualizers"—in his own discipline. As such,
he is frequently invited to participate in prestigious policy
and planning activities at national and international levels.
His opinions are sought and respected by the National Acad-
emy of Sciences and several influential intergovernmental
policy bodies.

    Scientist-politicians also exist within the narrower sense of
the term. Many scientists proclaim no political leanings, and
may disavow any association with political systems, yet the
more astute of their membership recognize the political
sources of power and funding, and metamorphose into mirror
images of politicians. Some undergo the metamorphosis skill-
fully, and interact successfully with their politician counterparts. Characteristics to look for include the following:

- Scientist-politicians enjoy personal contacts with politicians and their aides, recognizing the impressive but
  transient power that such people have over policy and
  funding, and identifying the influential players immediately. Furthermore, the successful scientist-politician
  is one whose face-to-face contacts with politicians lead
  to mutual respect, based on perceptions of mutual
  benefit.

Dr. Robert Aston has held a number of prestigious govern-
ment and university positions. Beginning his career as a
government research scientist and achieving some profes-
sional recognition, he then moved to a government-sup-
ported granting agency, then to the directorship of a new
national granting agency, then to a large university as vice

president for research, and then to the presidency of a college/university science consortium.

His job satisfactions are and have been many, but his role in shaping policies and in aiding the growth of a new national granting agency must have been the greatest. It is an opportunity afforded to only a few, and his performance, by general agreement, was exceptional.

Dr. Aston is particularly able in interactions with elected officials and legislators at state and national levels, and in communicating research needs and plans to decision-makers of all kinds. Over several decades he has developed a nationwide communication network of politicians, university heads, industrial leaders, and government bureaucrats, and has maintained these important contacts even through changes in his own scientific roles. He is adept at care and feeding of advisory committees and review boards, recognizing their potential utility to programs.

He is urbane, intelligent, and personable, very sensitive to the "people components" of any action. He has maintained throughout his career competency in his original discipline, to the point where he regularly offers a graduate-level summer course.

- Scientist-politicians are fully aware that political decisions are rarely if ever based wholly on scientific information. Such information may be used if it is expedient to do so and if it supports political objectives—otherwise it will be deemphasized or ignored. The decision process hinges on political–economic considerations and not on scientific evidence, even if the issue is basically a scientific one. When advantageous, research scientists can retreat easily to "lack of definite proof" for a given correlation (as in many current environmental controversies). Definite proof, as most of us recognize, is like absolute zero—theoretically possible but not characteristic of the real world. Scientist-politicians will take that one additional step and make inferences from the best available data, instead of withdrawing to a position requiring additional data before any conclusions can be reached.

- Scientist-politicians in the federal system recognize that their programs are part of a large trade-off pool in the hands of legislators or political appointees—to be used, bartered, expanded, contracted, or eliminated at the whim of the politician.

- There are many well-qualified scientist-politicians ready to assume key policy roles as politician-bureaucrats. Only a few are chosen (often by political recruiting teams of a new administration) for a moment of visibility and power—to disappear with a change in administration. During their reigns, the successful ones insert administration policies and objectives into activities of the career bureaucracy, and for that brief moment have great influence on the course of government science. Once deposed, most never reappear in positions of power, but some continue to influence political decisions as lobbyists or consultants.

Dr. Jonas Danforth represents a narrow category of scientists cum politicians, properly labeled politician-scientists, who launch themselves from a reasonable scientific base into various rarefied levels of politically supported administrative jobs. Their initial political associations may be results of accidents of family ties, or deliberately nurtured relationships with those holding or hoping to hold political power.

Dr. Danforth provides an excellent example of a career with heavily political overtones. With a good Ph.D. and several years of moderately successful research behind him, he emerged suddenly as a staff assistant to a key political figure in a federal agency. Before the party lost the following election, he was appointed director of a field research laboratory—normally a position not vulnerable to political manipulation. After several chaotic but not unproductive years, he accepted a university research appointment—to reappear at the end of that appointment as a special assistant to the head of another federal agency, presumably as the potential leader of a newly organized group within that agency.

Dr. Danforth has a brilliant mind, is a good thinker, and is capable of great charm as well as monumental intolerance and insensitivity. His survival abilities do not seem tied to any individual elected or appointed politician, but may be related to long-term family political associations. Whatever the base, he prospers, and because he is astute scientifically, he makes a positive contribution to policy and planning in science-related matters.

Dr. Danforth represents the ultimate politician-scientist, with maximum survival potential in a system that knows of his high political connections. He will appear and reappear in positions of power and authority, retreating to staff or university jobs in times when the party is not in full control. His self-evaluation must be one of success and achievement.

- Some scientist-politicians are diverted early in their professional careers as research scientists to become protégés of legislators or political appointees. They tend to be intelligent, vocal, and well-trained. Their tenure is, unfortunately, tenuous, rarely exceeding the time span of the administration in office at the moment, or that of the politician responsible for their appointment.
- Some scientist-politicians reach a point of unresolvable conflict between deeply held personal views and an official administration position which they are expected to

endorse and support actively. Such a crisis is often resolved by retreating to an industrial, consulting, or international job—signaling unwillingness to accept prostitution as a way of life.

Administration policies leading to recent reductions in federal resource and environmental programs have been supported by many scientist-politicians within the affected agencies. A few incumbents, however, have been unwilling to tolerate deliberate elimination of environmental programs. One such person is Dr. Gloria Jennings, a good science administrator with a genuine concern for resources and the environment. After a few months of lip service to policies with which she had no sympathy, she quietly resigned and withdrew to a temporary United Nations position. To many of those among the approving onlookers, this seemed to be an example of "triumph of principle," a rare but refreshing example of integrity in a climate of slippery values. They were naturally delighted when the UN system recognized her worth and moved her to a policy-level environmental position within that organization.

How Scientist-Politicians Maintain a Power Base

Scientist-politicians recognize that they are playing in a high-stakes game, with rules and devices unknown to or only dimly perceived by the average journeyman research scientist. Discussions with successful scientist-politicians have yielded some tidbits of relevant information about this complex game, mostly about behavior within the governmental system.

Appointment to a senior-level "politically sensitive" agency position requires development and acceptance of a critical list of operating principles, which may contain elements of the following:

- Always endorsing administration policy, stated as that, without implying strong agreement or disagreement with it
- Being able to explain administration policies rationally, even if they do not seem entirely rational (public support of policies or actions is crucial to the success of those policies or actions, and to that of the politician-bureaucrats who are responsible for implementing the policies)
- Being prepared to make reasoned attempts to insert professional viewpoints into internal discussions of policy matters, especially when flaws in policies are perceived
- Being prepared to absorb blame for errors at higher administrative levels if necessary, but not accepting blame for actions of other peer-level scientist-politicians
- Being prepared to accept responsibility for the sins of your key career bureaucrats, but not tolerating too many mistakes before transferring offenders to other jobs
- Protecting the administration, but observing with great diplomacy the fine line of ethical conduct
- Accepting the dictum that indiscretions, even minor ones, render the miscreant vulnerable to dismissal—if the indiscretion is publicized
- Remembering that every agency contains a cadre of informers—whistle-blowers—who can and will expose the marginal or illegal deeds of political appointees
- Being aware that there is always a gap in the armor—a vulnerable point in any career—where a determined legislator can eliminate a political appointee through such devices as subcommittee hearings, reports of an investigative staff, hostile review of files of the agency, or General Accounting Office review of actions taken or not taken

- Remembering that departure under pressure from a key political bureaucratic niche should be accomplished with grace and without rancor for the system or the administration—any other response will be viewed negatively as sour grapes
- Recognizing that the basic responsibility of a scientist-politician is to translate administration policies, attitudes, and whims into agency actions. Effective interaction with senior career bureaucrats is critical; this is usually a love–hate relationship, but it must be made to work.

Scientist-politicians function in the murky territory between politics and the practice of science. They attempt to influence decisions on behalf of their discipline, their agency, or their university. They unite with each other when necessary to present a common front against budget cuts or funding diversions to other public agencies. They may even be a positive force (however selfishly motivated) for new program initiatives or increased support in broad areas of research.

The apparent and disturbing insight is that science is increasingly political, largely because of dependence on public funds—hence on those who control those funds. A cadre of scientist-politicians is clearly required to translate, mediate, and influence the widely disparate entities of science and politics.

With the emergence of government in the past three decades as a principal funding source for research, science and the political process intersect repeatedly. Scientist-politicians often function as intermediaries or as direct participants in political decisions based in part on scientific information. Obvious dangers exist in intrusions of the political process too deeply into the affairs of science. A dynamic relationship must exist, however, to prevent excesses on either side. The scientist-politician can have critical impact on the status of this relationship—and is for that reason alone important to his more bench-oriented colleagues.

THE SCIENTIST-ENTREPRENEUR

A common stereotype of the scientist is that he or she is usually "ivory tower," with little or no interest in "business." Here and there, though, there are scientists who don't fit the mold, who are really scientist-businessmen or business-women, and who are separable from ordinary business people only by their scientific backgrounds. They are not abundant, but they are distinguishable, and some are successful, as measured by realization of career goals. They may operate private research institutes, technology-based production companies, or consulting companies.

These scientist-entrepreneurs have some common traits. They tend to be well-trained, energetic, aggressive, financially oriented, and managerially competent. Some have established reputations in science; many have come from postdoctoral positions in university faculties; most are intense, dedicated, intelligent professionals.

The Entrepreneur-Director

Among the interesting occupants of this category of scientists are the entrepreneurial directors of private research institutes. These are individuals with a dream—to create a research group as they would have it, free from too many governmental or university restrictions. The dream becomes reality with award of private or public funds. These institutes are usually small, frequently devoted to fundamental studies, and often in perpetually precarious financial condition. Locations of the institutes are sometimes remote but usually pleasant (and even exotic).

Directors and/or founders of the institutes are reminiscent of the small-business entrepreneur. They have a vision, and are energetic and capable enough to make it a reality. They assemble professional and support staffs, create facilities, and provide some erratic continuity in funding. The best of them recognize the importance of creating a pleasant, stimulating, productive environment, frequently in a rural location of their selection, which may be well removed from academic centers. They recognize too the necessity of maintaining a critical mass of expertise—conducive to retention of the best people. They also recognize that a particular breed of scientist will survive and produce under the conditions offered, whereas others will not. They accept that stagnation is a major evil in isolated locations, and overcompensate with liberal travel budgets, opportunities for training and refresher courses, and active seminar and visiting investigator programs. Few of these entrepreneurial directors expect to get rich, but they do appreciate the prestige and power of the job. A danger is that with disappearance of the original entrepreneurial director and founder, the organization may regress or stagnate; research funds may wither; and staff may begin bickering—unless a strong aggressive replacement is found.

The High-Technology Entrepreneur

The "new" scientist-entrepreneurs are the genetic modification experts—the core groups, offshoots, and splinters who have developed, taught, and used techniques of transfer of genetic material from one organism to another. The immediate promise is, of course, the mass production of substances like growth hormone now in short supply and expensive, or interferon, also in short supply. Less glamorous but useful products these companies are close to providing include a high-protein bacterial-based animal food and a vaccine for treatment of hoof-and-mouth disease in cattle. The longer-term promise of these enterprises is for life-improving modifications of awesome proportions—such as control or eradication of certain hereditary diseases.

The significant new component is that research scientists are forming, directing, and profiting from these ventures. They are not merely paid employees any more—they are in charge. Whether they will retain control by hiring the appropriate financial and managerial staff remains to be determined in many instances. Whether the bright promises of commercially useful products will be realized also remains to be seen. At present a principal problem seems to be in orderly expansion—in moving successfully from the research laboratory to the production facility.

Another new aspect of these technology-based ventures is that some universities and institutes are also getting into the act, sometimes forming private entrepreneurial ventures in concert with faculty members (and occasionally with pharmaceutical companies too) to exploit joint patents. In other instances, universities have allowed faculty members to form for-profit companies or institutes while retaining academic ties.

During the period 1980–1982 more than 100 small research-oriented companies were formed, many by scientists or groups of scientists. Some of these ventures have gone public; stocks sell well at first, but investors seem increasingly skeptical. Some of the companies with weak financing have merged or failed, but new ones crop up almost every day. Large pharmaceutical and chemical companies (some owned or directed by scientists) have started internal programs and/ or have invested in the smaller pioneering firms.

The whole process leads to reaffirmation of a vague hypothesis that scientists like money too—but they are not usually adept at accumulating it in large amounts. When a route to wealth is found, as it has been with so-called "gene splicing," everybody wants the lead and the profits. The normal pathways of science—free discussion of research projects, informal reviews of status of knowledge, peer examination of proposals—all shut down during an intensely competitive and secretive period when one group tries to exploit a promising approach and other groups do the same with different approaches. Secrecy is alien to normal science and difficult to maintain; leaks occur; key people defect to competing groups; graduate students leave to set up their own operations; and eventually channels of communication open again—in preparation for the next level of problems waiting to be solved. Throughout this repetitive process, small advances or genuine breakthroughs occur, and some groups succeed while others struggle to survive, or fail.

Scientific entrepreneurs have to be dazzled by the potential in this "genetic engineering" business, since it does promise a revolution in the pharmaceutical, chemical, and agricultural industries. Barnaby Feder, business analyst for the *New York Times*, stated (January 10, 1982), "sales are expected to reach tens of billions of dollars by the end of the decade. The impact by the year 2000 is sometimes compared to that of electronics."

The Consultant

Whether in biotechnology, aquaculture, power plant siting, agriculture land use, or other developing areas with a scientific base, there is a continuing need for advice from experts. Consulting firms formed by and/or employing specialists exist to examine and report on, for a price, almost any issue or problem or question of the moment. Consultants come in all dimensions; some are risk-taking professionals who have no other support or staff; more are affiliated with a small specialist group to do consulting in that specialty area; still more are part of larger commercial ventures with a resident staff of experts and a long list of available part-time people. The larger ventures frequently hire short-term staff for the duration of a contract, and release them at the end of the study period.

Consulting doesn't have to be an all-or-nothing occupation. Many university faculty members act as part-time consultants for industries or industry organizations. Today's demands are in molecular biology, and it is rare to talk to a specialist in this area who is not consulting or has not been approached for paid advice. Environmental consulting has also been popular during the past decade, coincident with public concerns about actions of polluting industries, and about any activity which degrades air, water, or land.

The extent to which universities permit or encourage faculty members to do outside consulting varies greatly, but most research institutions now have guidelines of some kind. Some, such as Harvard, Yale, and Stanford, allow one day a week for industrial consulting. Some require disclosure of consulting arrangements and financial interests. Problems of primary loyalty and commitment arise when faculty members are paid with stocks or stock options in the company; when they become officers in industrial concerns for which they have been consulting; or when they become the principal executive officers of private consulting ventures.

Professor Mortimer Bowen is my prize example of an excellent scientist cum consultant-entrepreneur. After a decade of effective research and teaching at a major university, and after successful management of several large federally funded multidisciplinary projects, he founded (and incorporated in another state) a private consulting venture to take advantage of the then-increasing environmental concerns, and their impacts on chemical and power companies. This was a time of regulatory actions requiring environmental impact studies before project approval. His company was available, it hired credible professionals, and it prospered. With a small permanent staff and with several field offices, its operations spread over much of the West Coast. A key to its success seemed to be a pool of part-time university faculty members and a much larger pool of postdocs, graduate students, and undergraduates hired for short-term work. Professor Bowen was an energetic, highly organized, gregarious man, recognized for his earlier research, and for his obvious commitment to the university as well as to his company. The university was tolerant (some would say too tolerant) but eventually he was asked to give up either his tenured position or his private business interests. He elected to sever his academic connections, and claims to this day that the decision was the correct one.

These, then, are a few examples of what must be called scientist-businessmen. They are often excellent professionals, but with an added dimension that involves them in high-risk ventures, for power or profit or both. Their backgrounds may be in research, but their principal commitments quickly become management oriented.

THE INTERNATIONAL SCIENTIST

The last category in this series on destinations for scientists, and to many the ultimate blend of all others—researcher, educator, administrator, bureaucrat, entrepreneur, and politician—is the "international scientist." Charismatic, elitist, professional—representatives of this special group epitomize much of what is best in science.

The group does not (repeat *not*) consist only of the "public" international scientists, the few who flit from continent to continent offering prophesies, dispensing advice on programs, or serving as activists for pet causes. For every public performer of this kind there are hundreds of less visible but excellent professionals who function internationally and who hold the fabric of international science together. They tend to cluster in and around large international organizations such as the Food and Agriculture Organization (FAO) or the World Health Organization as employees, consultants, or panel members. They are often involved in bilateral or multilateral scientific projects with developing countries.

Critical qualities of such professionals are political awareness and perception, expertise in bureaucratic maneuvers, and scientific credibility—as a specialist, a synthesizer and integrator, or as a planner and organizer of working groups and joint projects (with ability to let others lead when it is politically expedient to do so). Additional desirable qualities include multiple language abilities, a flexible family able to live abroad when necessary with "zest," ability to develop rapport and friendships with foreign counterparts, ability to cope alone with novel personal and research situations, and ability to stay interested in science affairs even if located in remote regions (principally through a preexisting network of colleagues).

Credibility as a scientist is of course paramount. Foreign scientists frequently have an excellent command of the world literature in their discipline, and they are perfectly capable of separating the credible from the phonies. It is important to be able to participate in working sessions with nationals of many countries with recognition as a scientist, and not as just another foreign bureaucrat.

Reasonable questions are "How does a scientist get on the international track?" and "How does he or she stay on that track?" Answers to the first include sheer chance, early and effective participation in an international meeting, application for a short- or long-term job with an agency such as the FAO (everything from a three-month midcareer stint in Zimbabwe or Agadir to a permanent post in Rome or Geneva), recommendation from a colleague already on the "inside" in an international agency, or none of the above. Answers to the second are simpler—excellent performance and/or assistance of properly placed friends and colleagues.

Innocents Abroad

Good international scientists can be recognized by *where* they travel and *how* they travel. One of the "best of the breed"—Professor Ernest Sodoma, formerly of the FAO—once described in detail what he had learned about foreign travel and attendance at foreign meetings. His narration was crowded with useful findings and advice—some just a trifle extreme, some with a soupçon of personal prejudice, some based on very inadequate sampling, but much with good insight and even a little humor. He has graciously permitted partial summarization of his comments here. They have been assembled under three headings—not mutually exclusive, and not totally inclusive either. They are (1) general guidelines, (2) meetings in selected industrialized non-Communist countries, and (3) meetings in eastern European countries. Categories (2) and (3) obviously leave out much of the world (all of Africa, for example) but meetings in the Third World countries are rare and are subject to special rules.

1. General Guidelines for Foreign Meeting Attendance

So much of the pleasure and profit from foreign meeting attendance depends on attitudes; active friendly involvement is a requisite, as it is at any meeting. Some admonitions can help to prepare the scientific voyager:

- He or she should be prepared for often interminable papers in languages which he or she will not understand.
- Simultaneous translation is the best way to ensure a degree of communication, but, because of exorbitant costs to the organizers, is usually not available. (For example, at a recent world conference with an attendance of 1100 people, the cost of three-language translation, with equipment, was $25,000 for five days. Such a price is financially destructive to any but the most substantially funded meetings, and is usually available only at those sponsored by intergovernmental agencies such as the United Nations.)
- If formal meetings in non-English-speaking countries are to serve any useful scientific purpose, oral translation services by professionals must be available despite the cost. The best plan in the absence of simultaneous translation is to present a short summary paper, with translation by an interpreter sentence by sentence, into the language of the host country, and to have copies of the complete paper available in English, since many scientists can read English but have great difficulty with oral presentations.
- The need for interpreters also exists at coffee breaks and in other informal discussions, since many foreign scientists are reluctant to carry on sustained conversations with colleagues who speak only English.
- Meeting plans and agendas are often "flexible," so participants should be prepared to present their paper at any point in the proceedings, to modify the length of their presentations drastically with little or no notice, or

to develop entirely new presentations to fit the last-minute needs of the organizers—and to do any or all of these things with aplomb and professionalism.

2. Meetings in Selected Industrialized Non-Communist Countries

Professor Sodoma has had opportunities during the past decade to participate in meetings in many countries, and related that some western European nations and Japan were great places for scientific meetings. Some of his observations for the record seem useful here:

- Probably nowhere in the world is scientific meeting attendance a greater pleasure than it is in the British Isles. Good scientists are abundant, scientific exchanges are sharp and stimulating, decorum is proper, exquisite low-key humor often abounds, and the whole proceeding is in English. Sodoma recently attended an oil pollution meeting at the Royal Society of London which included all those delightful qualities, and was held in a setting and city with charm and tradition. Being highly alert during the sessions is necessary, though, since discussions of papers can be tough-minded, penetrating, and carefully critical on occasion.
- Meetings in France have a special flavor too, once a few ground rules are learned. The French approach to science and to scientific discussion tends to be more leisurely; the French prefer to have the meeting conducted insofar as possible in the French language (which does not seem to be too unreasonable, provided all participants have some knowledge of the language); the French consider long evening meals as periods of relaxation and not as mere extensions of the day's meeting;

and the French are proud of the status of science in their
country.

- Meetings in West Germany, Belgium, or the Nether-
  lands will be conducted with thoroughness and precision. Meeting arrangements will be correct and inflexible, discussions of papers may be blunt and direct, and
  the weather can be dreadful. Good scientists in those
  countries may seem reserved, but they warm quickly
  when they recognize competence in foreign visitors.

- Meetings in Japan also have special and unique features.
  Paper presentation is formal; papers will often be read
  carefully, word by word, in preselected and carefully
  practiced English. Discussion of papers may be minimal
  or totally frustrating because of the language problem,
  and the Japanese too are quite reserved on initial contact
  with foreign scientists. The reserve disappears during
  the elaborate evening parties which are a feature of
  meetings in that country. Sake, solicitous "waitresses,"
  exquisitely presented but unusual food, and active participation by foreign guests in the singing and entertain-
  ment are all part of the enjoyable ritual. Field trips to
  historical and other places in this beautiful country
  should be planned as an integral part of meeting attendance; if at all possible, try to be accompained by local
  scientists.

3. Meetings in Eastern European Countries

On rare occasions, meetings of international scientific
organizations may be held in the eastern European countries
of the Communist bloc. Special guidelines for attending such
meetings were outlined by Professor Sodoma as follows:

- Don't go—unless you are part of an organized tour
  group, or unless you have an excellent command of the

language, or unless you are to be accompanied constantly (and that means constantly) by an interpreter or host representative. If none of these conditions are met, much of your time in eastern European countries will be spent in simply surviving—trying to understand strange signs, rules, and customs; trying to get from place to place; trying to buy a bottle of mineral water; trying to change an airline reservation—all leaving little time for scientific matters. It is just not sensible to try such a trip on your own if you hope to accomplish any scientific objectives.

- It cannot be stressed too strongly that some working knowledge of the language of a country should be an absolute prerequisite to a visit there. Dr. Sodoma pointed out that he and others had often proposed this to funding agency hierarchies, but were just as often ignored. His argument was that we in the United States persist in sending people abroad who are ill-prepared to interact with foreign counterparts in any language but English—which is a severe handicap.

- Go to meetings in the Soviet Union only with the written invitation of an agency official, and have a clear understanding that you will be "hand-carried" by a member of the host organization throughout the visit. The alternative, and a poor one, is to place yourself in the impersonal, thorough, and inflexible hands of the "Intourist" travel agency, and to do everything by the numbers. As long as you are accompanied constantly by a native, travel and living problems in the Soviet Union may not be overwhelming. If, however, plans are misunderstood, and you are left on your own in public places or on public transportation, you will encounter a harsh uncomprehending society with little time for or interest in your problems.

- Air travel in Communist bloc countries has unique and unlovable aspects which should be understood. "Confirmed reservations" is a meaningless term, and schedules fluctuate drastically and almost whimsically. Baggage handling is at best "casual" and exasperating waits are usual. Customs lines may be agonizingly slow and searches can be detailed (Sodoma described a half-hour of watching with horrified fascination, at the Moscow airport, the entire contents of a fellow passenger's suitcase being emptied as *each item* was placed *separately* in the X-ray machine). Three hours should be allowed for international check-in, and a transit schedule which allows less than three hours between planes should never be accepted from a travel agent. If connections are missed or travel plans must be changed, days of constant frustrating effort may be required.

- Just as is true in many places in the world, travelers in Communist bloc countries must be accompanied by a sense of humor—to survive the seemingly senseless rules and regulations, the deliberate disinterest of all categories of service people, and the discomfort of travel in cramped overbooked planes, trains, and buses. This does not imply that you should be crowded out, or ignored, or abused, however; some firmness, foot stamping, and desk pounding are required occasionally.

- If you are at all sensitive about lack of cleanliness in public restrooms, restaurants, and hotels, a visit to a Communist bloc country—the Soviet Union in particular—may provide interesting experiences. Service people are abundant, but as a group they seem to dislike their jobs intensely, and provide as little service as possible, as grudgingly and unpleasantly as possible. Most public toilets, even in academic institutions and museums, are abominably filthy; food in all but the most

expensive restaurants is served under marginally sani-
tary conditions; and hotel rooms, even those listed as
first class, are often dirty and poorly furnished (if judged
by United States standards).

- Here and there in Communist bloc countries you will
encounter good scientists and good institutes. They are
like gentle islands in a sea of pushing, uncaring, rude
humanity. If you feel that personal contact with such
scientists is worth the trauma, then you should go, fully
armed with the precautions already given.

- Probably the most important dividend from a visit to a
Communist bloc country is the establishment of some
personal rapport with its scientists, many of whom (particularly in the Soviet Union) have never visited the
United States or any other Western country. Future
communication and information exchanges can be substantially enhanced by even brief face-to-face encounters. Many of these people are remarkably well
informed through the published literature, but they are
often eager for more personal interaction with Western
scientists.

Professor Sodoma's concluding thought, and the
rationale for this exposition on travel, was that excellent
scientists act as professionals regardless of their surroundings, but anticipation of variables in a foreign
environment can increase the pleasures and rewards of
participation in international science.

The Scientific Jet Set

Hovering over all these good and substantive international scientific endeavors is a somewhat ephemeral minor
group that deserves grudging (and perhaps envious) mention.
This is the "scientific jet set"—scientists with power (technical,

administrative, or both) who move freely and frequently to exotic places to attend conferences, participate in elite invitation-only workshops, or review ongoing projects. They appear in the host location with a faithful assistant (usually female if the principal is a male), and sometimes with a retinue including a PR person and the laboratory "chef d'affaires." For his presentation, the principal draws on the productivity of his entire laboratory group (with scant or effusive praise for them) and interacts socially and professionally with counterparts from other scientifically advanced countries for a few brief hours or days. He then is off to meet other pressing obligations. The question is occasionally raised by onlookers (privately, of course), "When do these people teach or do research, in view of all their involvements elsewhere?" The usual answer is, "Rarely, but their assistants, teaching fellows, and postdoctoral fellows cope."

CONCLUSIONS

If this book has a core or nucleus, then this chapter on "Destinations for Scientists" should be it. In these pages scientists have been taken out of the laboratory, where many would choose to remain and where some do remain, into complex worlds of management, bureaucracy, and politics. They have been placed on planes, trains, and donkeys to travel the face of the planet—all in the interests of science.

These are all worthwhile evolutionary steps, given the structure of scientific research in these waning years of the 20th century. Science is today more than ever before an integral part of society—a force for good as well as for evil. Major decisions are made every day which are based in part on scientific information; it is certainly preferable that they be made by, or at least have the advice of, scientifically trained administrators, bureaucrats, and politicians.

The scientist, then, is far more than a laboratory-bound stereotype. His or her career destinations are diverse, interesting, and absorbing, but the base remains the same—productive, innovative, relevant research and teaching. Beyond this base, though, are career extensions of the kinds described in this chapter, which bring unique satisfactions and rewards. These extensions, whether they be managerial, political, or entrepreneurial, are most readily available to excellent scientists—those who have established credibility and have succeeded in the practice of science.

REFERENCE

1. Harriet Zuckerman, *Scientific Elite: Nobel Laureates in the United States* (New York: Free Press, 1979).

CHAPTER 7

THE ASCENDANT FEMALE SCIENTIST

Strengths and Weaknesses in the Female Armor; "Pushing the System"; Emerging Concepts in Risk Assessment and Risk Management for Female Professionals; the Female Science Manager; Sexuality in the Science Workplace—Some Realities and Fantasies

INTRODUCTION*

Dr. Millicent Curry, 39 years old and an associate professor of physiology and biochemistry, filed the last printout of the contaminant analysis, leaned back against the lab bench, and asked herself, "What in hell am I doing in this place at nine o'clock on a Sunday evening?" She knew the answer. She was

---

*In a previous book in this series *(Winning the Games Scientists Play)* I was able, despite almost universal objections and complaints from female scientists, to retain and to publish a chapter innocuously titled "Evaluating the Roles of Women and Men in Science" (previously titled "Sex in the Laboratory"). I am continuing, in the present volume, my dogged examination of the female scientist—again despite strongly stated advice against it from female acquaintances (none of whom admit to the classification of "friend" since the first book was published).

winding up a major piece of research which would result in another substantive publication in a series that had already brought her promotion and recognition by peers as a key figure in her chosen area of study.

She knew too that there were more and more women across the country who, like her, were getting a bigger piece of the action in science, by hard work, ability, and a changing climate for females in the scientific community. She also knew that her present faculty position was beyond anything she might have expected from the system as it existed even two decades ago.

Career sequences for female scientists seem to be (from the author's admittedly very limited perspective) less stable and less subject to neat categorization than are those of male scientists. This is probably true in large part because of the upheaval in and overturn of traditional subservient roles of females in science. Formerly, the perennially underpaid female assistant professor, held at the same rank for 20 years, was a predictable phenomenon on most campuses; but the recent insistence by women on equal career opportunities and rewards has created chaos in the prognostication business. We are now confronted with more and more women of all ages who insist on writing their own "scripts," or who reject traditional roles in science altogether. They want lead roles rather than subservient ones, and the numbers of female scientists achieving this goal are increasing.

FEMALES WHO ARE MAKING IT

It should be instructive to examine briefly the areas where women are clearly in the ascendancy—even beyond the accelerated upward movement which characterized women in most of science. Some identifiable areas are (1) the upper echelons of science-based federal regulatory agencies (such as the EPA and FDA), (2) the middle bureaucracy of government granting agencies, (3) government research laboratories, and (4) scientific communications (in all media).

Partly as a consequence of continued insistence by successive political administrations on the placement of women in key positions, an impressive number of impressive females with science backgrounds now occupy management and policy-level positions in the hierarchies of many federal agencies, especially the regulatory organizations. Some of them are obviously political appointees, whose tenure, according to the established spoils system, will not exceed that of the present administration; others are good career bureaucrats who have moved through the system quickly by a combination of competence, assertiveness, and a favorable national climate for upward mobility of females.

The midlevel bureaucracies of science funding agencies, public and private, are also well populated by females, many of whom have excellent scientific credentials. Part of the explanation offered (usually by male administrators) is that women have perceptive and analytic capabilities equal to or better than men, and that they are adept at grasping quickly the complex and critical interpersonal components of the job.

Good female scientists are finally making it into senior positions—as measured by salary and support staff—in federal research laboratories. Part of this movement is clearly a spin-off from the persistent government campaign of upward mobility for women, but most must be overdue recognition of competence and productivity. The author has visited laboratories recently where half or more of the key research/supervisory positions were occupied by females. Such a situation would have been highly unusual even a decade ago.

Another area where women with science backgrounds are ascendant is in the broad range of activities called "communications"—from television science reporters to editors of science-oriented magazines and bulletins. The science community has recognized, somewhat belatedly, that effective translation of complex research and its rationale into reasonably good, clear, interesting prose is a prerequisite for adequate financial support of research. Often in the past, translation responsibilities have been left to (and have been shirked by) individual scientists, but more and more—especially in the past decade—people with science backgrounds and an overlay of communications skills have been inserted between the research scientist and the public. This has been true especially for some of the government-funded programs such as Sea Grant. Substantial numbers of these new specialists are female; they can now be found in managerial as well as operational positions.

Any reasonable search would undoubtedly expose additional scientific or science-related areas where females are moving up—as measured by relative increases in numbers and responsibilities. Free-lance and technical writing is one such area. Another, which defies an old female stereotype about difficulty with quantitative concepts, is the broad field of computer applications in science.

STRENGTHS IN THE FEMALE ARMOR

Somewhere early in this discussion of the female role in science we should reemphasize the truism that any final judgment of a scientist, regardless of sex, will be based on individual contributions to knowledge, and not on skill at sexual or other kinds of game playing.* With the gradual movement toward equality (despite some lingering biases of a traditionally conservative and male-dominated occupation) it is apparent that the female role in science has inherent strengths as well as weaknesses. The female power base is supported by several basic premises: (1) female scientists must still survive and prosper in a male-dominated conservative system; (2) the upwardly mobile female scientist will seek out and use to a maximum extent the sources of power in that system; (3) power is still concentrated in a *hierarchy* of male supervisors, managers, deans, and executives, rather than in a *single* key power figure; and (4) there is a greater appreciation today than ever before of the fact that the female scientist is equally as creative and productive as her male counterpart.

The importance of social interactions in science, as adjuncts to the traditional discipline-oriented productive activ-

---

*The fascinating topic of "Sex in the Laboratory" was explored (or at least confronted crudely and unsuccessfully) in my earlier book, *Winning the Games Scientists Play*. Irate responses (mostly from females) have encouraged further probes into this surprisingly sensitive area. Good but ego-deflating insights have been gained from uniformly hostile correspondence and more direct forms of criticism. Probably the best insight was that science and sex are totally incompatible, and should never be discussed simultaneously in a single book, especially by an obvious amateur. Another good insight was that all laboratory scientists, male or female, are really robots, not programmed to respond in any way to hormonal stimulation. Still another insight was that males in science have totally erroneous perceptions of females in science— that whatever those perceptions are, they are wrong. All of this leads to the ultimate conclusion: there does seem to be room for perceptive insightful consideration of male–female roles and relationships in science—but probably not by men. An author who is male cannot treat the subject of female inequality in science, or the matter of sexuality in science lightly or flippantly—in fact, any attempt at humor about these matters will be misconstrued or denounced as chauvinism. A male author cannot treat the vestiges of sexism in science with humor, since to most females there is no humor in the subject. Similarly, there is little margin for levity in the disparate treatment of females in the history of science.

ities, was emphasized in an earlier chapter of this book. Any serious study of how female scientists can succeed at interpersonal strategies which help to ensure a full and rewarding career reveals the importance of being noticed—of emerging from a white coat and tile wall background. Emergence can be enhanced by conscious cultivation of key people and power figures (still mostly male), and playing to their self-perceptions and vanities. (Male scientists can and do, of course, stand out from the crowd in similar ways.)

These and other kinds of direct personal exchanges obviously require perception, research, and conscious effort, but can be accomplished openly and sincerely by approaches such as inviting opinions and comments on proposals or manuscripts, by developing good but nonsexual relationships with male peers, and by encouraging appropriate mentor relationships with senior colleagues. It seems important too, in an area of continuing transition, for females in science to be always consciously "on camera," projecting real qualities of intelligence, confidence, and energy—since they are usually scrutinized more carefully than men.

From a strictly male perspective, it is obvious that women must insert themselves into traditionally male networks, from which they are often excluded at present. Entry is possible and frequently unexpectedly easy, with application of reasonable pressure. Participation in male networks is an important route to advancement, since members know more about the organization of science and about the events that give visibility, or enhance opportunities for new jobs and new projects. Coalitions and networks among female scientists are important and worthy objectives, as was outlined in a recent article, "Affirmative Action: Total Systems Change" by Alice G. Sargent,[1] but major effort should be spent on getting into the male networks, since that is where the power is concentrated.

Control and elimination of sexual harassment in the science workplace are also worthy objectives, which have received attention, especially in the past decade. Some guidelines include rejecting overt advances with firmness and action. Longer-term objectives should include the enforcement of laws and regulations which will discourage sexual harassment on the job.

WEAKNESSES IN THE FEMALE ARMOR

Even though the ascendancy of women in science is an observable fact, there are still pockets of persistent conflict and dysfunction—places where male–female interactions are less than perfect:

- Many female scientists still feel isolated in a male-dominated occupation. A sociological study of a decade ago, "Women in American Science" by H. Zuckerman and J. Cole,[2] disclosed several interesting aspects of this phenomenon. The traditional isolation has not disappeared concomitant with upward movement. Women still do research alone or with other women, and it is more likely that two women or two men will publish together, but rarely a mix of these. It is also more likely that good female faculty people will attract female graduate students, and will publish jointly with those students.
- Women have not been given entry in significant numbers into the still male-dominated informal scientific "clubs" discussed in Chapter 4—groups which control much of the substantive action within particular subdisciplines. Until they do gain full partnership in these cliques, they will remain as vocal support components of science, but not integral units. This lack of acceptance

into informal power groups was suggested in another recent sociological study, "Women in Science" by J. R. Cole,[3] as a possible major deterrent to "full scientific citizenship" for women.

- The roles that females play in science are, in the opinion of a number of women who have been interviewed as part of the research on which this book is based, forced on them, usually in graduate school or earlier, by men and a male-dominated system. Many females have been raised and educated as *females* (in the old traditional sense of noncompetitiveness in team activities, accepting male–female societal roles, and even tolerating male domination) and some continue to react accordingly in professional situations. For them, training in science does little to change these basic response patterns. Today, however, there is less acceptance of male dominance, but until recently most females had little choice except to leave the profession.

- Some of the old complaints persist: "Women in science are not taken seriously by men"; "Men want to talk to other men"; or "Women are a welcome diversion, but are not expected to lead or dominate." Prejudices of this kind take a long time to disappear completely.

- A persistent problem still exists for women in isolated, high-stress, often uncomfortable field science positions. The author has observed some of the substance of the problem on research ships taking long cruises. The presence of female scientists imposes a whole new set of rules of conduct and decorum on both males and females. Male scientists break the rules occasionally in cases of sexual harassment. (This is not to say that similar problems have not occurred during field research on land as well.)

- Older female scientists are still being misused by the system.

A good example of persistent discrimination in a changing society is Dr. Harriet R. Traverse, wife of the eminent Dr. George B. Traverse, and a good scientist in her own right, with a Ph.D. and a long distinguished history of teaching at a major university. With her husband's last move five years ago to a privately funded research institute, she was offered a desk and nonsalaried "cooperator" position so she could work on a peer basis with her spouse (her area of expertise is similar to his)—an offer which she accepted on a temporary basis, but she is still in the position. A more assertive and younger woman would have been less likely to accept or tolerate this peripheral role. The important fact to note here is that such arrangements do exist today, despite all the clamor about equality.

"PUSHING THE SYSTEM"

The new era of equal rights for women carries with it a heady atmosphere which encourages the thesis that anything is possible if pursued aggressively. Often, though, an entrenched conservative system (of which the science establishment is an excellent example) will resist change which is construed as too rapid. Assertive females must "push the system" to correct long-standing inequities. Usually they win, but not easily and not without some foot-dragging by the power structure. Female scientists want equal access to training programs, to professional meetings, to foreign travel, to promotions—to all the "perks" and rewards of a professional occupation. This new assertiveness is good, if supported by high productivity and excellence on the job; otherwise it may encounter at least passive resistance from supervisors and administrators. (Some females would claim that passive resistance exists regardless of productivity.) Provision of rewards to any staff member, regardless of sex, is in the end a matter of subjective judgment by the power figure, and correct appraisal of the supervisor's response is a critical consideration. Requesting too much, too often, too fast can result in negative decisions, bureaucratic delaying tactics, and even hostility from colleagues as well as supervisors.

Examples of the "too much too fast" syndrome are not difficult to identify. Consider the case of Elisa Macintosh, an intelligent aggressive female journeyman scientist. Her on-the-job performance, which involved extensive computer interaction, was good, and she has served as chairperson of the laboratory EEO Committee. For the past three years she was given released time to take job-related courses at a local university, even though her employment was only part-time (at her own request, because of family commitments). She recently approached her supervisor with a request for additional course work aimed at a Ph.D. degree—these courses to occupy more than half of her paid duty hours. At this point her supervisor rebelled, based on prospects of further diminished productivity in the essential job to which she was assigned. He took a strong stand against any increase in released time activities, insisting that general degree courses were individual responsibilities and must be taken outside normal duty hours. This stand produced a complaint to the EEO Counselor, which was subsequently escalated to a regional level. No decision has been reached yet, but she may win. Some of her male peers are already quietly envious of what appears to be favored treatment by "the system."

"RISK ASSESSMENT" AND "RISK MANAGEMENT" FOR FEMALE SCIENTISTS

Stumbling around interminably in the dangerous (for a male) jungle of opinions and conclusions about women in science has resulted in the capture of at least one important concept—that of the importance of "risk assessment" and "risk management." "Risk assessment" is an attempt to quantify risks which exist as a consequence of an action or modification of an action. It is an attempt to express the probability of specific outcomes due to specific causes, and its elements include identifying hazards, estimating risks, evaluating social consequences, and comparing estimated risks with others. "Risk management" can be defined as strategies determined to be necessary in dealing with risks and in rendering them neutral or acceptable. These terms are of course general ones in the insurance business and, more and more, in other areas. They normally include risk analysis, protection against catastrophic losses by spreading responsibilities, careful calculation of the likelihood of success of any venture, preparation of fallback options in the event of initial failure, and construction of actuarial tables for predicting disappearance of projects. Risk management is especially critical during the vulnerable early phases of a scientific career, but for female scientists its importance appears to continue indefinitely.

The initial risk assessment in early career phases is probably most critical, and requires painstaking analysis of all available data. In the matter of developing a professional relationship with a mentor, for example, consideration must be given to perceptions of a male mentor's awareness of and susceptibility to male–female differences and similarities, and to any tendencies that he may have toward unrealistic fantasies. In another example—professional relationships with male colleagues—female scientists should examine the most appropriate form of day-to-day interactions, particularly the degree of formality or informality to be established and maintained. Frank discussion of these matters with male colleagues is desirable, but often avoided until problems develop. The consequence of inadequate risk assessment and management here may be distortion or truncation of what might otherwise have been a useful professional association. Female scientists must evaluate with great thoroughness the risks and advantages of too-close personal involvements with males within the organization—a topic with delightful as well as dangerous components, and a topic worthy of book-length treatment (but not here).

THE FEMALE SCIENCE MANAGER

Women are appearing more and more frequently as managers of scientific organizations. The author has conducted extensive surveys of this emergent female—particularly in the waning hours of cocktail parties at scientific meetings, where certain moderately inflammatory opinions have provoked ridicule and threats of physical violence. It seems difficult or impossible for a male author to sit comfortably on a fence and to view managerial women dispassionately. He immediately metamorphoses into a chauvinistic pig or a patronizer—with a thin and very sharp division between the two, and acres of thorny underbrush on either side. During this painful learning process, a series of simple dicta for male scientists has been developed, and is offered here for guidance in coping with the managerial woman:

- Many women are very good at what they do.
- Many women are better than men at what they do.

- Women are gradually emerging in scientific management in significant numbers, and they are and will be supervisors of men.
- When women assume managerial responsibilities, they bring to the job all the virtues and ills that characterize male managers; some are perceptive, decisive, open, and honest, and some are not.
- Women (as well as men) are found in upper managerial strata because they are exceptional, or because they have traveled part of the way with the guidance and support of an executive superstar—or (as is often the case) by a combination of the above.
- The successful female managers are those who can participate effectively in a male-dominated system. They (the female professionals) ask, and expect, but they give too, as professionals. They expect support from those in authority for projects and requests, but they in turn provide support for others—often more junior colleagues.

All is not sweetness and light, though. Exhaustive research has further indicated that the principal problem with female managers is that they must deal with stupid precast men—a surprising proportion of whom still have rigid concepts about the proper role of females. This negative factor may decline and disappear as a new generation of males and females moves into science management, but it is still apparent today. Also, it seems that excessive assertiveness on the part of a minority of women contributes significantly to whatever problems exist.

Many of today's male science managers have had neither extensive work experience nor friendships with managerial females. They grew up in science when females were technicians, and never appeared in managerial roles. It is small won-

der, therefore, that they are ill-at-ease and unprepared for the
new attitudes and demands of female professionals.

Female managers differ from male counterparts in fundamental ways, as was pointed out by A. G. Sargent in her 1981
article cited earlier in this chapter. Her general findings were
that women tend to

- Use formal organizational processes more consistently,
  whereas males use informal power systems
- Place higher value on intrinsic rewards, such as self-
  esteem and achieving personal goals, whereas males
  value the extrinsic (salary, promotion, visible signs of
  power)
- More frequently react rather than take initiatives in
  group discussions
- Are more "production oriented" than males. (Other
  studies have also concluded that the female manager is
  more inclined to be production and quality oriented,
  whereas the male manager tends to be preoccupied with
  other job factors which ensure survival and comfort.)

It is important for males to recognize these and other differences in evaluating competence, and in interacting successfully
with female managers. (It is important also for males to recognize that mere enumeration of these differences may be the
basis for a new way of stereotyping female science managers.)

SEXUALITY IN THE SCIENCE WORKPLACE

The role of sexuality in male–female interactions in the
work environment, scientific or otherwise, is a subject which
evokes strong opinions whenever it is approached. A. G. Sargent, in her perceptive article discussed in the previous sec-

tions, summarized some interesting observations on the matter as follows:

- Sexuality is always present in male–female interactions.
- Sex is and will continue to be a pervasive issue in the work environment.
- Men and women almost always appraise one another sexually (even though this has nothing to do—or should have nothing to do—with professional appraisal).
- Sexual relationships, real or fantasized, can develop through work contacts.

These axioms of Sargent are as relevant to the science work environment as to any other, and should be recognized as such. Under some rational restraint and control, gender differences in the science work force can be positive factors in sustained productivity and maximum use of talents. Without control, sex can be a disruptive and destructive force, in science or elsewhere.

Problems created by mixing sex and science are astoundingly variable in nature and intensity. Most of them center on diminished productivity by distracted, bemused, sometimes irrational professionals and assistants—whose total attention may be focused temporarily on their libidos, to the exclusion of any other responsibilities or interests. The range of possible sexual interactions in a permissive scientific environment comes very close to overpowering even inveterate creators of artificial categories. Some common threads exist, though, and these may be exploited in this discussion:

- Despite admonitions and appeals to common sense, overt or clandestine sexual relationships develop within scientific organizations, frequently between an employee and his or her supervisor.

- Sexual relationships between two members of a scientific group lead almost inevitably to the disappearance from the group of one or the other of the participants.
- If one participant is a female staff member in a relatively junior position, she is usually the one targeted for departure or transfer.
- Sexual activity between female supervisors and male assistants is not unknown, but is infrequent when compared with its reciprocal. This may be in part because the female supervisor is intent on making it in a male-dominated system, and sexual dalliance is a luxury to be dispensed with and a clear deterrent to upward movement in that system.
- Homosexual activity—even by inference—between two members of a scientific group is unacceptable. Scientists still tend to be conservative, even reactionary, in their views and judgments on personal lives of colleagues.
- Intralaboratory sexual relationships are, more often than not, intense but short-lived. If the termination of the relationship is less than friendly or mutual, the work environment may become for a time a hostile place—for onlookers as well as participants.
- Many staff members welcome the opportunity to be gleeful spectators at the rise, the full heat, and the decay of intralaboratory sex episodes—especially if they violate marriage vows.
- Some staff members believe that suspect or abnormal behavior on the part of a colleague in his or her personal life reflects lack of trustworthiness in scientific matters. Thus, flagrant violation of accepted sexual mores could be considered as evidence that the professional ethics of the culprit(s) may be questionable as well.

Clearly, behavior associated with gender differences is a not-insignificant aspect of on-the-job conduct within the ivory tower. Management of sexual activity is not a usual course offering in scientific or supervisory training, but maybe it should be.

CONCLUSIONS*

Progress is apparent in developing equal status for women and men in science. Movement is too slow for some and too fast for others (roughly and unevenly divided on the basis of sex of the respondent). Erosion of stereotypes is leading gradually to creation of an entirely different set of roles in the science workplace. Despite some persistent resistance from a male-dominated system, women are occupying senior scientific and managerial postions in increasing numbers.

Successful female scientists—successful scientists in general—must have as part of their power base a superior record of professional productivity. Superimposed on this can be the numerous interpersonal activities which characterize many excellent scientists of either sex. Removal of former barriers to

---

*A logical question, as this chapter ends, is, "Why make such a big thing about differences between female and male scientists?" Maybe the perceived differences are imaginary or trivial compared to the similarities—or maybe it is simply that the differences are more interesting to read about than the similarities. Whatever the reason, the matter seemed (and still seems) worth exploring—but probably more sensibly by a female.

Another logical question is, "Where's the joy?"—and, considering the title of the book, the question is not an unreasonable one. The answer to it must be, of course, that female professionals experience the same highs and satisfactions in the practice of science as males, so the content of other chapters applies equally to both sexes. It is only in the added frustrations imposed by a still unbalanced system that differences exist.

upward mobility has resulted in emergence of more and more females who represent excellent combinations of the productive scientist and the aware individual.

The female role in science can be characterized by the existence of strengths and weaknesses. Strengths include recognition of a persistent institutional problem of attitudes, which has led to more direct confrontation of the problem, with salutary results in terms of reduced discriminatory or sexist actions. Weaknesses include persistent exclusion from the informal "clubs" which dominate subdisciplines, and the persistent slowness with which women achieve full scientific citizenship—as measured by rank, salary, and authority.

REFERENCES

1. Alice G. Sargent, Affirmative action: Total systems change, in: *Consultations on the Affirmative Actions of the U.S. Commission on Civil Rights* (Washington, D.C., 1981), pp. 147–158.
2. Harriet Zuckerman and J. R. Cole, Women in American science, *Minerva* 13 (1975), pp. 82–102.
3. J. R. Cole, Women in science, *American Scientist* 69 (1981), pp. 385–391.

CHAPTER 8

AUTHORITY FIGURES IN SCIENCE

The Dual Sources of Authority in Science; Uses and Abuses of Managerial Authority; Science-Based Authority—Moving Gently but Carrying a Big Stick; Multiple Rewards and Punishments of Mentoring; The Inevitable Indispensible Chairperson

INTRODUCTION

Professor Arthur G. Hoechst finished his morning lecture to undergraduates in Biochemistry 102, drove quickly to the airport, and in two hours was careening by taxi through the rainswept streets of Washington—on his way to a critical meeting with his NSF Grant Program Director. As coordinator of a large multi-institutional grant, Dr. Hoechst was to learn today if a new and much larger Phase Two proposal was to be funded by the agency. The review process has been extensive, with both on-site inspections and many outside reviewers. This was to be the moment of decision.

Professor Hoechst reflected during the brief ride from the airport on the confrontations of power and authority that were involved in the day's activities. After all, he was the manager of a research group of 17 people (exclusive of graduate students), and coordinator of the larger grant-supported program, which included research groups from five institutions in widely separated locations. He was meeting the Project Manager and the Program Director of NSF, who were responsible for evaluation and final funding approval of his proposal. Yet the chairperson and members of the site visit team, not to mention the many anonymous reviewers of the grant proposal, were probably the critical voices in reaching a decision.

The interplay of roles—advisory and decision-making, managerial and professional—constitutes a basic mechanism in the practice of science in the United States, and it works. "Authority" has been, is, and will continue to be an elusive concept in science and elsewhere. It has been defined in many ways, but seems in its simplest form to mean having the power to control one's own actions and those of others. Authority in science is somewhat unique in that it is derived from two quite different sources—managerial authority (usually appointive) on one hand, and scientific authority (based on scientific accomplishments) on the other. Uniqueness is seen in that the two sources of authority need overlap little if at all. Commonality is found in that both kinds may have profound effects on the scientific careers of others.

This chapter explores briefly some highly selected aspects of this dual base of authority in science—managerial and professional—and then scrutinizes in greater detail two other less clearly defined authority bases—the mentor and the chairperson.

MANAGERIAL/ADMINISTRATIVE AUTHORITY FIGURES

Many successful scientists are also authority figures, in that their administrative activities and decisions can and do affect the careers of other scientists. However, the authority figure in govenment/industry research is not necessarily a successful scientist in the sense that we have been using the term "success" in this book. He or she may have a political or managerial base for authority, and yet have only marginal scientific competence. To achieve some order in this analysis it might be worthwhile to isolate, discuss, and dismiss these managerial authority figures first, and then to consider authority figures whose base is scientific competence—accepting the inevitability that there may be a small zone of overlap.

Management of research organizations is more complicated than that of most production groups. The help is discriminating and demanding; product quality-control criteria are somewhat subjective; and the products themselves (scientific papers, patents, books) are often difficult to evaluate.

Some scientific organizations are managed well and some aren't. Those that are have people in charge who can have drastically different management styles and philosophies—but they must be effective in maintaining or encouraging research output. Scrutiny of management styles can be a fascinating hobby, but it leads to frustration if the observer likes to develop neat categories and to insert people into them. Management styles can be as varied as the practitioners—which is not surprising, since each manager brings to the job a unique mix of background, attitudes, and experience. There are, however, some characteristics of government/industry scientific managers which emerge from the case histories examined:

- They usually have great and multiple enthusiasms and above-average intelligence.
- They can channel and focus those enthusiasms so they peak consecutively, with a new crest rising as the preceding one wanes.
- They are great at subtle orchestration of events. They can envision steps in a process far in advance of most others.

- They are without exception highly perceptive.
- The best are forceful, even overpowering, so their point of view prevails, even in the presence of good alternatives.
- They may be very credible as scientists (except in most cases of political appointments).

In addition to these positive attributes, some government/ industry science administrators exhibit certain negative Machiavellian tendencies in their exercise of authority. These can also be as varied as the people involved. At a recent three-day seminar for science managers, one of the lecturers, Dr. Robert A. Hilton, former director of a federal research agency, discussed in detail some negative approaches to management. Making it clear that he neither condoned nor advocated any of them, he developed a partial list of negative strategies sometimes employed by science managers:

- Experimental psychologists have concluded that "positive reinforcement" is a better motivational force than fear. Despite this, some managers still subscribe to the concept that full use of authority requires the *evolution of fear*—through a progressive series of steps or the repetition of fear-inducing methods (sarcasm, cutoffs, putdowns, derogatory comments) until the individual target and the group are mute, subservient, and responsive to any request.
- A *crisis environment* can be perpetuated by science managers—so that all participants in any project or activity will operate at maximum adrenaline levels continuously.
- *Self-reinforcement*—the enhancement of the authority figure's visibility and perceived importance to the organization by assimilation of the efforts of others—is a primary and totally justifiable use of strength.

- *A well-thought-out game plan* is essential for every manager, but it should not be revealed too early. An air of uncertainty about approaches, plans, and policies can be perpetuated (through such devices as hidden agendas not disclosed to others in meetings until they start).
- Any action which promotes a *sense of dependency* on the part of subordinates should be considered.
- Authority figures select good effective midlevel people for short-term or acting roles, let them do the work, give them adequate rewards or punishments, but never let them forget who has final authority.
- Some administrators create a complex matrix organization where *lines of authority are blurred or obscured*, except that all lines lead to the central figure.

This list of negative strategies is obviously too short; any scientist who has survived under or worked in close proximity to such administrators could add to it substantively. Items which could be included are the "favored few" syndrome, in which certain projects always seem to be recipients of increased funding and additional staff; the "kitchen cabinet" group of key advisors (which may not include all of the principal operating managers); the "pet project" of the authority figure which receives funds and staff diverted from authorized projects—the list itself could easily become book length (but not now).

The bottom line in the exercise of managerial authority in science is of course productivity and (secondarily) creativity. Whether these objectives are achieved by humane participative management approaches or by less desirable techniques matters little in the final objective evaluation of a project, laboratory, department, or institute. Scientists (some of them at least) can survive and even prosper under radically different kinds of regimes, from permissive to dictatorial. Some scientists, however, can prosper only under a narrower range of management philosophies; these people migrate, or their productivity declines, when the management environment shifts too drastically.

Dr. Ivan Fortunoff was a laboratory director in the traditional European manner—an autocrat in complete control, with a highly selected staff of submissive but productive scientists. Entrepreneurial administrators of his type dominated research in United States in earlier decades, but have largely (though not entirely) disappeared. All the research planning emanated from their offices, and scientists were often treated as high-grade technicians. In the example at hand, Dr. Fortunoff's laboratory was known for several decades as one of the most productive in the system (which included 16 other laboratories). His staff consisted of survivors of a severe winnowing process—survivors who could adapt to an authoritarian regime and retain productivity. That laboratory probably represented an extreme, though, in which the director's name appeared on almost every publication, and in which the staff could be assembled in clean white lab coats at their laboratory doors in a moment's notice, to make a brief presentation on their research to visitors.

The Loss of Authority

Exploration of the consequences of *loss of authority* would be an appropriate postscript to any discussion of managerial/administrative control in science. Authority has been recognized to be a consequence of brains and ability, combined with chance selection from a large field of equal or superior competitors. It has been further identified in some instances as the consequence of a political decision. Some of the pedestals of authority are therefore extremely fragile, and the withdrawal of any one may result in collapse of the entire structure. Random unpredictable events such as a change in political administration, a major agency reorganization, loss of a principal funding source, or active disfavor of a key decision-maker at a higher administrative level, may result in diminution or disappearance—quickly or gradually—of managerial authority. A position may be abolished, or combined with another; an entire management level may be eliminated or reduced drastically; or an incumbent may be simply replaced by a vote of the board of directors, a decision by the operating executive, or a political maneuver.

The sense of loss and deprivation that accompanies loss of managerial authority can be severe. If not a part of the normal career expectations of a scientist-administrator (and it probably should be), the loss of control can be a devastating event, from which some individuals never recover. Often, though, the displaced administrator has several fallback options (a temporary staff position elsewhere, a consulting job, or a reentry into teaching or research).

The "ripple effects" of loss of authority can sometimes match the collapse of a house of cards. When a key figure disappears, all the fabric of his or her organization can rip off; sycophants immediately become vulnerable; and general unease settles over the entire group. The former authority figure usually leaves, or is banished to the lower reaches of the new organization, to become nearly invisible. Former key aides and lieutenants are usually replaced, programs are reoriented, the table of organization is revised, and the previous management policies and philosophies are disavowed.

Dr. George Hamilton was an autocratic, demanding, but effective administrator of a regionally dispersed federal agri-

cultural research organization. His several multidisciplinary laboratories carried out research in support of industry, and in response to needs of management groups, as well as basic science. Recently, he elected to give up his position, possibly with some pressure from the central office. He withdrew to a minimum-visibility position of staff assistant to the agency director, but remained in his original field location. A new administrator with a drastically different management philosophy was appointed, and after an appropriate learning period, reorganized some of the major research teams, displacing a few program leaders and elevating others to principal operating positions.

The former authority figure, now ensconced in a new office in the basement, remote from the administrative section, has been discreetly silent about any new management actions, and has withdrawn entirely from program interactions, professing complete contentment with resumption of a research career.

SCIENCE-BASED AUTHORITY FIGURES

In an earlier chapter on "External Signs of Success," the subjects of "networking" and "the club" were explored. Networking and becoming part of the club are approaches used by competent scientists to acquire and use authority. Nonmanagerial control in science resides principally in leaders of "in-groups" within any narrow specialty, in journal editors and referees, and in regular participants in permanent advisory groups to granting agencies. This kind of authority is less obvious than that of the manager/administrator, but is equally important to the careers of scientists. The obvious and somewhat unique reason for this is that *scientists depend on evalua-*

tions, opinions, and actions of colleagues external to the organization for which they work. Participation in "in-group" networking assures that the members will be aware of projects or findings far in advance of information in the published literature. Favorable reviews and acceptance of manuscripts by editors and referees of major journals assure publication of the kind that is meaningful to peers and to faculty evaluation committees. Membership on advisory boards reporting to policy levels of government agencies or foundations assures that one's opinions will be heard and that colleagues will value those opinions.

Science-based authority is diffuse and relatively inconspicuous, if compared with managerial control, but it is at least of equal importance in determining success or failure in professional careers. The "authority figure" and his or her retinue present at a scientific society meeting, the remote demanding journal editor, the anonymous and often equally demanding manuscript referee, the anonymous clique of colleagues who meet regularly to offer advice to granting agencies about which proposals to fund, the colleagues who discuss in corridor conversations the merits and deficiencies of published papers and books—these are the nonorganizational arbiters of success in science.

THE MENTOR AS AN AUTHORITY FIGURE

Transfer of the essential components of science—tradition, history, ethics, approaches to thinking—from the older to the newer generation is a critical part of the continuity of the discipline. It is done most effectively through master–apprentice relationships, the masters being mentors for their bright, aggressive successors. The typical and stereotypic association is the professor–graduate student one, in which the accumulated wisdom and experience of the senior member is transmitted over an extended period to the junior member. In examining case histories of successful scientists, the crucial and quietly powerful role of the mentor emerges again and again. Since this is a factor which is often underemphasized in graduate education, particularly in larger universities, it seems important to affirm its significance here.

Mentors come in assorted sizes, shapes, and sexes; their function is to serve as father (or mother) figures, role models, inspirational sources, door openers, motivators, confidants, and of course teachers. Mentors need not be senior people (though they often are); on occasion they may be peers or even (exceptionally) junior colleagues, with whom a special rapport develops. Occasionally a succession of mentors may characterize a career, with each one in sequence making an impact on it—some greater than others.

The typical mentor–apprentice relationship serves many purposes. Mentors

- Demonstrate a style and technique of doing research—serving as a role model
- Inculcate an analytical approach to selection of significant problems and development of appropriate methods to solve them
- Convey a method of considering, analyzing, and criticizing in professional constructive ways the work and conclusions of others
- Discuss in great detail the fundamental concepts in any subdiscipline, and the history of development of those concepts
- Offer an example of how to teach the next generation
- Illustrate the techniques and the value of informal networks in science

- Transmit by example and discussion the best methods of scientific writing
- Provide adequate time for informal discussion of scientific issues of the moment

A critical but often poorly identified function of mentors is *imprinting*—imparting a way of thinking and acting that distinguishes the professional from the amateur, and demonstrating the correctness and utility of that way of thinking. Imprinting is a complex and continuing process, best approached by example. Some principal components of the imprinting process are

- Attempts to define the boundaries of ethical behavior in the practice of science—and to reduce the subjective interpretation of those boundaries by major professors, fellow graduate students, ad hoc committees of professional societies, journal editors, and (occasionally) offended colleagues
- Acceptance of the often harsh process of "peer review" of manuscripts submitted for publication in professional journals, with its subjective, ego-deflating, sometimes destructive components—but with its very necessary critical evaluation of the value of contributions to the literature

Imprinting is a principal and essential function of graduate schools, equaling in importance the imparting of the conceptual and factual background of a professional discipline in science. It results in a perspective, a mind-set, a rational approach to the universe that is permanent and fundamental for a career in science.

Imprinting occurs in so many ways—day-to-day exposure to the graduate faculty and especially the thesis advisor, discussions with faculty members and other graduate students, brief contacts with seminar speakers, observations made at scientific meetings, and editorials in scientific journals. Some graduate schools have attempted to formalize the imprinting process through courses with euphemistic labels such as "professional ethics," "the philosophy of science," or "the scientist and society." These can be instructive, but the basic conditioning must come from "on-the-job" experiences—and the mentor should be at the core of many of these experiences.

The mentor relationship, carefully nourished, can be one of the most rewarding and profitable personal interactions in science—for the donor as well as the recipient. Some mentor relationships develop well beyond the master–apprentice level. They may persist for years, and may involve collaboration in research and writing, as well as joint participation in conferences (where the junior colleague is introduced to the senior's peers and to those of potential importance to the junior's future career). Usually, though, there is a sensed time limit, after which the junior must emerge as an independent investigator, building on the earlier relationship but not bound to it.

Satisfactions to the mentor are many: proper preparation of a new generation of professionals, pleasure in providing motivation and background for emergence of exceptional practitioners, and the quiet joy of constant association with intelligent highly motivated individuals. There are, however, some negative aspects. Whereas many mentor–apprentice relationships are based on respect, admiration, even love, it is not unusual for some to lead to ambivalence, jealousy, and conflict. Mentors can be possessive, opinionated, and autocratic; apprentices can be stubborn, aggressive, and ungrateful. Accommodations and compromises are usually required, as in any interpersonal venture.

The intrusion of sexuality cannot be discounted if the mentor and apprentice are of different genders. Male scientists who are mentors for younger female scientists often face the problem of limits to a professional relationship. Female mentors for younger male scientists can face the equally complex problem of avoiding the appearance or actuality of a sexual relationship. Solutions are varied, and not always satisfactory to both parties.

The male–female donor–recipient relationship in science is becoming more common, as women make up larger percentages of graduate student populations, and as women move into science management positions. Male mentors in such alliances must accept additional guidelines and strictures, since the aura of sexuality is almost always present—whether perceived or actual. Persistent attention must be paid to impressions of the nature of the relationship gained by colleagues and casual onlookers. Even the most straightforward and professional of relationships may be misunderstood or misinterpreted by some.

Examination of the careers of successful female scientists or science administrators, as well as published results of surveys of women in business, suggests that success in many cases is linked to a long-term professional relationship with a male mentor (since there are far fewer women in high positions) who is a leader in that field. The rise of the female member is tied to the rise of the male. Sometimes too the female selects a series of mentors rather than a single one; each provides a degree of leverage for success.

The potentially profound effect of a mentor on the career of a scientist cannot be overemphasized. Future job sequences can be altered or even determined by some small facet of the mentor–apprentice relationship. A true mentor, after all, will make known the significant contributions of his or her apprentice. Actions by the mentor, often of unrecognized importance at the moment, can open or close doors to career options. What greater power—and responsibility—exists?

THE CHAIRPERSON AS AN AUTHORITY FIGURE

The mentor discussed in the previous section functions on a one-to-one basis with protégés—an indispensible *personal* relationship in science which has its counterpart in the *group* relationship headed by a chairperson (or convenor). The group may be a committee, a subcommittee, a department, a working group, a review panel, an advisory board, a working party, or a task force—but whatever its designation, a principal and inevitable element will be its leader, the "chairperson." Some chair positions require a single performance—such as a panel discussion—whereas others extend over months or years, and require additional skills. Because of the ubiquitousness of group activities within and on the periphery of science, this focus of the authority of the group deserves careful attention— just as careful as that given to the mentor in the previous section.

Chairpeople constitute an elite subclass of professionals whose members appear consistently as leaders of discussions, planning group meetings, small group conferences, scientific meeting sessions—any situation where several to many people assemble for a stated purpose. Effective conduct in the chair-person role is partially innate and partially an art form whose principles can be learned. Although the job is usually by appointment or election, the same effective people are selected repeatedly. Watching good chairpeople at their work can be a source of pleasure and wonderment for the informed observer, and a source of enlightenment for the novice.

"How to chair a meeting" has been the subject of many books and manuals, and the broader aspects of group dynam-ics—most of which involve a leadership figure—have been dealt with *ad nauseum* in psychological journals and books, as well as in an endless stream of "how-to" and "self-help" paperbacks. So what is new or unique about leadership of scientific groups? In view of the plethora of available literature, what possible contribution can be made by a few pages in a book on "The Joy of Science"? These questions are admittedly daunting but not overwhelming. It turns out that there are concepts and operating procedures which are particularly relevant to the scientist/chairperson, and that they seem important to any discussion of the exercise of authority in science.

A propensity for developing partial lists is, as may have been noted with some frustration, an expanding and possibly unlovable characteristic of this book. A rationalization is that such lists, to be really complete, would be interminable. A more honest reason is sheer laziness on the part of the author. In any case, a complete list of principles of scientific chairpersonship, if double spaced, would probably stretch from the session room to the society president's suite. Well hidden in that list would be the pearls and nuggets for which we are searching. What follows is the result of one person's selection.

- Chairpeople—whether short-term or long-term—come in assorted flavors, bringing to the role a variety of attitudes and techniques. Some favorites are

  —The "dictator"—stern inflexible, no-nonsense, and correct (if not always right)
  —The "Rodney Dangerfield" who invites abuse, uses sarcasm freely, and encourages its return
  —The "Milquetoast"—an apologist who rarely makes an unassisted decision and searches interminably (but usually successfully) for consensus
  —The "great compromiser" who takes no hard stands and makes no flat statements, but who may be extraordinarily successful in developing consensus.

The assumption of any of these roles, if done with intelligence and with the right kind of group, can be successful in achieving high productivity. Most groups appreciate a clearly understood style of chairpersonship, and can tolerate almost anything except a bland dull performance by the chair.

- Scientific chairpeople often exemplify a poorly explored concept of "clustering," which has implications far beyond committee activities. According to the concept, a single individual holds an entire group together—whether they be committee members, working group participants, or academic department members. Any change in his or her status or energy input will change or threaten the structure of the entire group. Additionally, the productivity of the group is shaped to a large extent by the chairperson.

- Scientific chairpeople learn quickly that to achieve the best results, there is no substitute for sharp, relevant comments from the chair, laced with touches of harmless humor, taking into account the feelings and opinions of others. They learn just as quickly to suppress any tendency to dominate discussions or to push a particular viewpoint too hard.

- Scientific chairpeople also learn quickly that if an impasse is reached, if tempers seem to be rising, or if a crisis of any kind seems to be developing, excellent responses include calling for an extended coffee break, or, *in extremis*, retreating to the rest room and leaving the meeting in chaos. Often during such intervals a certain magic (accompanied by some earnest small-group discussions) will produce a resolution for almost any problem.

- Experienced scientific chairpeople resist with great firmness any proposal to draft or edit a document during a

committee meeting. Absolutely nothing is more stultifying or more wasteful of professional talent than line-
by-line construction or modification of any document.
This work should be done by individuals or small
subgroups, and presented for approval to the entire
group.

- The role of the scientific chairperson includes the full
expectation of lonely evening hours preparing draft
summaries and recommendations from the disorganized
records of the day's activities. The informed chairperson
does not, however, expect the committee or panel members to participate in evening sessions—even informal
ones. After a day of intense personal interaction, they
mostly want to think of other things and do other things
(such as visiting nightclubs, watching an HBO movie on
television, or reading a newspaper). If liquid refresh-
ments are provided (or even if they are not), most evening gatherings quickly become social rather than scientific events—which is entirely proper, but a little
disconcerting to the overly aggressive chairperson.

Other maxims might be listed, but these examples should
be enough to provide a few insights. Convenors and chairpeople, then, are crucial to the many group activities important to
scientific organizations and to scientific progress in general.
Successful conduct in the role, as measured by productivity of
the group, can be a source of great satisfaction (yes, even joy)
to those who lead and to those who participate.

CONCLUSIONS

"Authority" in scientific organizations has its base in
appointment to an administrative job (managerial authority) or

in professional accomplishment (science-based authority). Of the two, managerial authority is more visible and more controversial. Scientific managerial authority is a consequence of intelligence and ability, combined with chance selection from a large field of equal or superior competitors. Management styles are as varied as are the individual managers, but there are many commonalities in the management of scientific organizations and in that of other types of production organizations. Scientists respect and respond positively to properly applied authority; scientists expect that they or their representatives will play key roles in major decisions; and scientists recognize that whoever controls funding must also control programs.

Some of the sources of authority in science are, however, outside the organization and independent of it (as are some of the rewards). This schism provides roles for authority figures whose base is accomplishment in science. As examples, research leaders in any discipline or subdiscipline control certain staff appointments and selection of graduate students, journal editors and referees control the acceptance of manuscripts, mentors control (in part) the destinies of their protégés, and chairpeople control the composition and productivity of committees and working groups.

Such a dispersion of authority can create tensions within research organizations, but can also provide alternate routes upward for those excellent scientists who want to lead and not to follow.

PART THREE

DESCENT FROM CAREER PEAKS

The second major section of this book moved with some measured exuberance over the crests of scientific careers, pausing briefly to examine a careful selection of upbeat episodes. That short odyssey may have encouraged a sense of uneasiness about so much joy—a vague feeling of incompleteness, that down in the valleys, always in the shadows of the peaks, there must be a rougher, less satisfying terrain that many follow. That is the counterpoint, the reality, for a significant proportion of scientists. It would be less than honest to ignore or minimize the low points, even in a book with the grandiose title of *The Joy of Science.*

We come then, however unwillingly, to the flip sides of the hit records—the persistent mediocrity, the problem areas, the infirmities, the dissatisfactions—that are also part of science. Exploration of the pits may expose some reasons for lack of success; this may be a redeeming factor justifying the diversion and helping to guide with greater realism the search for excellence. The choices for what may be a somber trip include such dismal topics as controversies and frauds in science, burnout and fade-out, guaranteed losers, the aging scientist, and the futile search for immortality—all part of a presumably necessary effort for a balanced presentation, even though the book title promises a clear bias toward the positive.

CHAPTER 9

THE PATHOLOGY OF SCIENCE
Controversies and Frauds

One of the Few Universal Truths in Science: People Fight, Sometimes Dirty; Major and Minor League Controversies; Frauds—the True Pathologies of Science

INTRODUCTION

Some of the material in this chapter has been excerpted (with permission of the executors of his estate) from the brilliant and no-doubt classic series "Lectures on the Pathology of Science," presented seven years ago at the Harvard Outpatient Clinic by the late Professor Murchison C. Krumwalter, Distinguished Professor of Tropical Public Health. In his lectures, Professor Krumwalter drew analogies between the abnormal in science and his experiences in the practice of medicine. For example, he compared minor controversies in science with the daily office encounters of the general practitioner, and major controversies with genuinely life-threatening illnesses. Later in the series he discussed obvious frauds in science with the same clinical detachment afforded to inhibitants of locked wards of

181

mental hospitals—but with less professional sympathy for the abnormal scientists.

Professor Krumwalter offered few new insights about therapy for scientific disorders, but his analogies were so vivid that this shortcoming was easily overlooked. Drawing heavily on his material, it seems relevant and necessary in this book on "The Joy of Science" to review the subject of controversies and frauds in science—principally to identify the boundaries of expected ethical practices, but also to point out that even exceptional scientists may be involved in controversies during their careers.

Some of the sources and effects of controversies will be examined, drawn insofar as possible from actual episodes. The spectrum ranges from minor disagreements about authorship of joint papers to accusations of deliberate falsification or theft of data. At some poorly defined point in the spectrum, proper scientific ethics come into serious question. Perceptions about the location of that point vary significantly among participants, onlookers, and arbitrators—hence the problem.

CONTROVERSIES

Controversies have characterized every era in the history of science; leaders as well as relative unknowns have partici- pated, often with great vigor and vituperation. Such contro- versies have been powerful attractants for determined and fas- cinated analysts. Merton,[1] in a perceptive series of papers published in the 1960s, extracted much nectar from three cen- turies of scientific controversies, beginning with Galileo. Mer- ton's comments on this topic should be required reading for any serious student of the subject. His general conclusions were that most scientists consider their ideas to be their own "intellectual property," and they react violently and sometimes almost irrationally in matters of priority of discovery. Furthermore, their associates often react with even greater anger. He cited many examples of simultaneous scientific discoveries and the ensuing controversies over priority. The winners in these scientific shouting matches are remembered; the also-rans disappear and are forgotten in a few generations.

A noteworthy fact is that similar ideas and the vague recognition of principles frequently occur almost simultaneously to more than one scientist. The fuzzy origins of ideas and concepts in science are recurrent problems. Principal credit should go rightly to the one who not only first enunciates the principle or concept, but who also explores and substantiates a factual basis for it, and who brings the information to the attention of colleagues and the world.

Recently, William Broad and Nicholas Wade, in their book *Betrayers of the Truth*,[2] slogged unhappily and even with some horror over much of the same ground as Merton, emphasizing many instances of fraud, some well known from newspaper and journal exposure, but others less well known. The authors concluded, probably correctly, that deceit and corruption occur in science as in any other human undertaking; they also concluded, probably incorrectly, that fraud is a common feature of the scientific landscape. Broad and Wade, unfortunately, seem to have missed several important points altogether—especially the obvious one that *scientists seem to enjoy controversies*, as long as they are not principals in the fight, and also that, after the dust has settled and the blood has coagulated, *controversies can and often do have salutary effects on scientific progress*. Controversies provoke careful reexamination of the statistical validity of data, and often force reassessment of conclusions and interpretations of those data. Frequently, too, the resolution of controversies clarifies concepts and leads to development of entirely new lines of inquiry.

Most good productive scientists, sooner or later in their careers, regardless of how carefully or delicately they tread, become involved in controversy—either as direct participants or as front-row spectators. The controversy may be relatively minor, such as disagreement over the interpretation of data, or it may take the form of a more severe reputation-destroying kind. Whether based on actual or perceived misdemeanors or high crimes, controversies damage the image of science as a calm dispassionate search for verification of natural laws. Exposed, if only fleetingly, are some of the elements which make up the hairy unattractive buttocks of science.

Problems are created by differing definitions and interpretations of the nature of science and the nature of scientific ethics. Because discrepancies exist, it seems worthwhile, early in this chapter, to review very briefly some descriptions of what science is all about, and to examine some of the perimeters of scientific ethics. Then, maybe, once the boundaries are established, we can speak freely about abuses and excesses.

Science has been defined succinctly by Stephen Schneider and Lynne Morton in their book *The Primordial Bond*[3] as "the interpretation of reasonably tested theories by frequently repeated observations of increasing precision, *always short of exactitude*" (italics mine). Other definitions abound in the literature; most of them include some elements of a search for truth or proof—for verification of general laws by specific observations. Insistence on verification separates the "hard" from the "speculative" sciences, but both are mixtures of inductive and deductive reasoning processes. Multiple difficulties can develop from differing perceptions of the adequacy of data, the justification for conclusions reached, or the validity of experimental approaches. Drastically divergent views of the nature of scientific evidence, if strongly held, can and do result in public controversies.

Ethical practices are even more difficult to circumscribe. "Ethics" has been described (unsatisfyingly) as "standards of behavior or principles of conduct governing an individual or a profession" or as "a group of moral principles or set of values." Ethical practices can then be described as "being in accord with approved standards of behavior or a socially or professionally accepted code" or "conforming to professionally endorsed principles or practices." Such descriptions leave much to *individual interpretations*—and it is the interpretations that can be and have been sources of controversies. It is rare for scientists to even think much about the ethical foundations of their profession until a controversy develops, and it is even rarer for any two scientists to agree on precise ethical boundaries.

Examining the spectrum of disagreements, it is possible to discern "major league" and "minor league" controversies. Major league status must be granted, regardless of the scientific importance of the issue involved, if the controversy

- Is reputation-destroying or career-damaging
- Is escalating to public forums or news media
- Results in legal action
- Leads to a request for an opinion by the National Academy of Sciences or some comparable body
- Includes claims of falsification of data or other forms of deliberate fraud
- Includes claims of wrongful use of another's unpublished data
- Includes claims of usurpation of another's concepts based on oral presentations of the other
- Includes allegations of other serious ethical violation

Major controversies are intimately interwoven with the onrush of science; every decade produces its own classic occurrences, as well as many others of a more mundane variety.

Minor or "bush league" controversies can have one or more of the following elements:

- Opposing interpretations of available data
- Claims of inadequacy of data supporting conclusions reached
- Allusions to naivete or unfamiliarity with the literature
- Claims of unauthorized use or misuse of data developed cooperatively by two or more individuals or groups
- "Unconscious plagiarism" of ideas or concepts discussed informally
- Simultaneous publication of parallel research results by two independent groups—each claiming priority
- Claims of unwarranted usurpation of senior authorship or sole authorship in publishing cooperative research results

Additional problems can surface during preparation of books or reviews:

- Accidental or deliberate failure to discuss or cite certain relevant publications
- Incorrect summarizations or conclusions attributed to published work of others
- Paraphrasing from publications of others without acknowledgment or citation

Unlike the major league controversies, the minor ones, which may for the moment seem larger than life, rarely lead to lasting professional damage, although they may encourage lifelong personal enmities and sometimes furious public arguments.

"He (She) published a paper using my data and ideas without giving me adequate credit." Variants of this pronouncement account for many if not most of the controversies which have been examined as part of the background for this chapter. A

female graduate student was put to work analyzing charts from a radio telescope designed by her professor (and employer). During their joint work she discovered signals from a new kind of star. She was given co-authorship on the journal article written by her mentor; he was awarded a Nobel prize; she went public with her version of the discovery.

A first-year graduate student, hired by a professor as a technician, was assigned to a field station where routine daily observations of air and water pollution levels were made. The student was given time to develop a related research project of his own, which he did. Unfortunately, the individual project depended on the data base being developed by the professor's grant. When the professor called for the routine data for analysis prior to publication, the student became very possessive about "his" data, and complained to the dean about "exploitation."

These and countless other unpleasantries which diminish the joy of science are based on *lack of communication* about roles, rewards, responsibilities, and priorities. Relationships develop and are modified as projects continue; the original agreements and guidelines fade, until credit for the work is about to be established by publication or public announcement of findings. At that critical point misconceptions about roles or erroneous perceptions about rights and responsibilities which have developed with time combine to form the basis for controversies—often ugly and usually damaging to all members of a research group.

A good and worldly associate, Dr. Frank Howe, who was involved early in his career with several such minor flaps based on inadequate communication, once told me about his final solution, which freed him forever from petty but not inconsequential disagreements with graduate students and other categories of workers making up a research team. The device was simple and well known to production managers, but apparently not to supervisors of research groups. It was a "terms of employment" statement signed jointly on the first day by Frank and his new assistant (whether technician, graduate student, or postdoc) documenting straightforwardly the specifics that otherwise could cause later misunderstandings. The new team member either signed or did not join the group. The statement was a classic in its completeness. I can recall only some of the categories, but the concept *was beautiful. Included were*

- PUBLICATION RIGHTS. *This critical section consisted of several typed paragraphs detailing Frank's policy on the publication rights associated with the particular job. If Frank was to be first author on every published paper from the group, this was so stated. If the employee could expect eventually to be a first author on a paper, the terms under which this would be done were also stated, in great detail. If the employee was to be included in a group of co-authors, this too was stated and defined. If the employee was never to be more than a name in the acknowledgments section of papers, this too was made known.*
- PAID HOURS OF EMPLOYMENT. *If a specified daily work schedule was expected, it was stated, together with something on credit time for hours outside the schedule, if allowed.*
- RESPONSIBILITIES. *This important catchall category included statements about recording data, any role in collating or analyzing data, briefing of visitors, administrators, or review groups, reporting anoma-*

lous results, caring for pet laboratory plants, picking up visitors at the airport, etc.

- RELEASED TIME FOR PROFESSIONAL ADVANCEMENT. *If the employee was to work full-time, any paid time to be allowed away from the job for training, academic course work, or short courses was stated clearly.*
- MEETING ATTENDANCE. *If the employee was to attend seminars or local meetings, this was stated. If the employee would be funded to attend professional society meetings, this would also be stated. If field trip travel was expected, the frequency, nature, and mode of such travel were stated. If that great plum, international travel, was to be feasible, the research requirements and frequency of such travel were stated.*
- SEXUAL ACTIVITY WITHIN THE GROUP. *Frank, a member of a vanishing breed, was a "hardliner" on this one, because of earlier traumatic experiences. No one, but no one, in the group was ever to get overly physical with any other group member. Friends yes; lovers, never. Departures from this dictum, adequately documented, resulted in automatic dismissal.*
- GROUP SEMINARS. *Every member of the group was expected to attend, and to participate actively in, weekly meetings of the group, which were mixtures of informal presentations by outsiders, reviews of the status of particular projects, or brainstorming for new projects.*
- EVALUATIONS. *Frank agreed to provide a face-to-face evaluation of each group member every six months, without fail. The reality was not quite that, but he did try.*

*The list went on, and seemed to get longer as each new
employee was hired—but each person knew exactly what the
terms were. In the 11 years in which I was familiar with
Frank's group (until he took an administrative position with
the National Bureau of Standards), no controversies
surfaced.*

"Unconscious plagiarism" (the counterpart of "cryptam-
nesia" described by Robert Merton in an article, "The Ambivalence of Scientists"[4]) is a source of much unhappiness in science, and is worthy of additional attention here. Plagiarism *per
se* is clearly definable as unauthorized or uncredited use of
another's published work, but "unconscious plagiarism" can
be considered by some scientists to extend to the use of anything heard in individual or group discussions—chance
remarks, joking comments, or brief oral summaries of work in
progress. There is some question, however, about how far it is
ethically necessary to extend credit for chance remarks or comments. Some scientists are very possessive (sometimes almost
paranoid) about their thoughts, ideas, and data; others are less
so, but always with some sensitivity. It should never be
assumed that a colleague will not be sensitive or even hyper-
sensitive about his or her scientific "possessions," even if he or
she seems relaxed about other things.

A relatively new area for development of controversies
between or among scientists is that of *adversary situations*—in
courtrooms, public hearings, or congressional hearings—
where professionals may be pitted against one another in
attempts to win cases or influence decisions. This harsh inhos-
pitable environment, dominated by shrewd predators with
legal training, is far removed from familiar scientific forums. It
is best exemplified today in medical, public health, and environmental areas, where conflicting conclusions, interpretations, or opinions may be offered from the same data base. The

scientist can be a pawn, and can even be damaged professionally, unless he or she is forewarned and prepared.

Adversary proceedings are especially troublesome for scientists because of the intervention of a middleman—the lawyer—whose objective is to win cases regardless of where truth may be, and whose world view is totally different from that of the scientist. Lawyers can and will destroy hard-earned scientific reputations with absolute impunity, if doing so increases the likelihood of winning a case. Expert witnesses with opposing points of view or differing interpretations of data can be very vulnerable, especially if they stray beyond (or allow themselves to be pushed beyond) the legitimate confines of those data.

Professor Nelson Marshall (real name) of the University of Rhode Island has explored the adversary problem in a thoughtful journal article "Advisor, Advocate and Adversary."[5] He emphasized the proper role of scientists as advisors with occasional assumption of an advocate role in environmental issues such as protection of tidal marshes. The adversary role to him is foreign to science and a source of many of the most serious charges concerning professional ethics (often resulting from commitment to an extreme position under the stress of cross-examination). He does conclude, though, that service in adversary hearings is a "legitimate pursuit in the outreach of a scientist's interests," if the pitfalls are assiduously avoided.

All this does not imply that differences in conclusions drawn from data should not exist. A very common and often healthy type of controversy develops from differing interpretations of the same or different data sets relating to a specific scientific question. If conducted properly, *within professional forums,* such controversies can be salutary—often leading to more careful analyses or to acquisition of more complete data. Sometimes, though, the disagreement escapes into the news media, where unsympathetic treatment by insensitive reporters can provide a spot of tarnish for all concerned, and for science in general.

An interesting but unproven observation is that scientists *enjoy* controversies, if they are not direct participants (and even on occasion if they are). As a spectator sport, observing the development, climax, and denouement of an episode captures the interest of professionals. Opportunity is provided to examine ethical positions, and to witness the testing of a system which is heavily biased toward the angels.

The startling (to me) realization that scientists may actually enjoy controversies occurred a decade ago in a confrontation over priority of publication. Several research groups had worked actively on a severe disease of terrestrial crabs. The outbreak had depopulated principal producing areas. The pathogen was thought to belong to an obscure protozoan group, but correct identification depended on recognition of the spore stage in the life cycle. Several research groups had seen stages tentatively identified as spores, but were reluctant to publish. One group, a federal laboratory, announced plans to publish its findings, stating that it had seen transitional stages as well as the spores. Response from the other groups was instantaneous. A second group, of university researchers (one of whose members had made the original discovery of the vegetative stage of the pathogen), announced that it too had seen the spores; that it too had a manuscript ready for publication (seven years after the original discovery); and that its identification of the organism should have priority. In the game of choosing sides, many of the other involved groups (all university researchers) sided with the second group (the university researchers). Com-

plaints of ethical violations were made to the director of the federal agency to which the laboratory belonged, and to anyone in the scientific community who would listen. Memos and accusatory letters flew like snow, and all informal communication ceased. A hearing was arranged in one of those huge, elaborate, mahogany-paneled Washington conference rooms—with row upon row of sober-faced formally dressed university scientists facing four representatives of the federal laboratory. That was the precise moment when I realized "These people are <u>enjoying</u> this event." It became apparent after brief questioning that there was no case for ethical violation to be made, and that all parties should sit down like sensible adults and work out a compromise, which would involve simultaneous publication of both papers. This was done; the controversy faded; and the opponents even speak occasionally to one another now. The episode did seem to have a salutary effect on science, though; it released a flood of papers from all the laboratories, many of whom had been reluctant to risk any accusation of violating the priorities of others—for seven years!

Unfortunately, and unlike other kinds of contact sports events, there are usually no undisputed winners or losers, except in cases of obvious and demonstrable fraud. Everyone implicated in controversies is injured to some degree. Even the spectators suffer, since they are diverted from the conduct of science in their own laboratories (even though they may consider the diversion to be worth it). Principal participants are often engrossed totally in preparing evidence supporting their respective positions, to the detriment of any other work in progress. Spouses and families are ignored, or are forced to suffer with the completely distracted and irritable participants. Friends and colleagues become enmeshed in the tumult as enthusiastic claques or as co-conspirators, and students are disenchanted to find that scientists can be just as mean, petty, possessive, and vindictive as anyone else. The important positive component of all this is that a legitimate cause or issue may be rightfully publicized.

FRAUD

Not all controversies are based on fraud, but claims of fraud can produce the most damaging of controversies, until guilt or innocence is determined. Controversies are universal; fraud is anomalous and, in the opinion of many scientists, infrequent in its proven prevalence among professionals. A concern expressed quietly, though, is that the true extent of relatively minor fraudulent activity in science is not brought to the attention of colleagues or the public. Strong differences of opinion about this obviously exist, and the problem is complicated by uncertainty in defining the ethical lapses which could be called fraud.

Some observers of science and scientists—for example Broad and Wade in their book mentioned earlier, *Betrayers of the Truth*—have been justifiably appalled by the perceived high incidence of cheating, lying, and manipulation of data that goes on in science. Such acts often result from unreasonable pressures to establish priority, to be recognized, to be promoted—and from the seeming ease with which small lies are accepted (even condoned) by the scientific community. These deliberate ethical lapses can involve the superstars of science—the so-called "world-class scientists"—as well as the average or mediocre. Simply to maintain position in a highly competitive environment can lead to stresses on ethical boundaries.

The problem obviously can't be solved here; what we can do is to create compartments for major and minor fraud (as was done for controversies), and then look at some possible motivations.

Major fraud in science involves deliberate acts which seriously subvert the concept that scientists search objectively for truth. Such acts may include the mortal sins of

- Theft and publication of data belonging to another
- Creation of entire data sets without any factual basis
- Falsifying or discarding data sets to ensure conformance with preconceived conclusions
- Theft and publication of manuscripts or portions of manuscripts belonging to others
- Clear and deliberate theft of ideas and concepts expressed by others orally or in manuscript
- Public announcement of a new method or a new synthesis based on manipulated experiments
- Plagiarizing the published work of others

These overt and deliberate instances of fraud, if proven, should raise serious questions about the future place of the perpetrator in scientific research, and should be grounds for disciplinary action by an employing organization. Unfortunately, in addition to these obviously reprehensible categories, there are murky areas near ethical boundaries, where decisions about fraud are harder to reach. Some relatively minor aberrations in this zone are

- Selectivity in getting samples or data which support a given conclusion or position
- Suppression of a few anomalous data points
- Selection of a statistical analysis which is supportive of a given conclusion, to the exclusion of others
- Performing extensive transformations and reductions of data sets to make inconclusive data appear conclusive

Hard-liners would reject any artificial distinctions between major and minor crimes; they would define most of these minor infractions as fraud, and they would not separate them from examples of major crimes against science. Other arbiters might condemn such practices as being marginally ethical, but might not propose a death sentence too quickly. A logical conclusion could well be that if science is really a search for truth, dependent on objectivity and honesty of its practitioners, then there is no room for lies or deceits, regardless of their magnitude.

Whenever fraud is discussed in scientific circles or exposed in news media, an inevitable question is "Why?" What motivates a segment of a professional population that is presumably bound by laws of objectivity to move outside accepted ethical boundaries? Clearly there is great individual variability, but motivation for major fraud may arise from

- An unconscious death wish—irrational and pathological
- A compulsive gambler instinct
- An overwhelming (also pathological) ambition to gain recognition by whatever means
- A conviction that the system has loopholes, and that no one will do the required follow-up
- A disdain for the value systems of science
- A rejection of the concept of "obedience to the unenforceable" in science
- A mistaken sense of intellectual superiority over colleagues, to the extent that they are considered too stupid to detect a deceit
- In special cases, belief in the immunity of the "authority figure" from prosecution by "ordinary" scientists
- Overriding greed for power and material gain
- Temporary insanity or severe emotional disturbance

- The intense competitive climate of current scientific practice, especially in the most dynamic research areas, in which success may be more prized than ethics
- Overwork, accompanied by inattention to fraudulent activities of junior colleagues or graduate students, because of pressures of other professional activities
- Pressures to have future grant applications accepted, based on interim results

Many perpetrators of major fraud state that *they honestly do not know or understand their motivation,* but then blame pressures of the job or the system.

Possible motivations for minor fraud—particularly in categories such as smoothing or otherwise manipulating data—may be a little easier to enumerate:

- Ambition to succeed quickly in a highly competitive profession
- The present permissive attitude of society—which fosters the concept that "a little lie won't hurt you"
- Knowledge of the looseness of control mechanisms in science
- Examples of colleagues who cheat and succeed

For broader roots to the causes of minor fraud, some observers like Broad and Wade who "view with alarm" the present situation, emphasize "the breakdown during the past few decades of the relationship between master and apprentice." They see a "dehumanization" of this relationship, finding it increasingly based on material needs, rather than on shared intellectual and personal bonds. This breakdown contributes to a more casual view of the ethical ideals of science. Some investigators, in a frenzy for money to support their graduate students and junior associates, as well as to purchase expensive equipment, will go to extreme measures. Other critics point to the increasing intrusion of political or bureaucratic objectives into science as another reason for minor frauds—the push to get adequate data quickly, the necessity to reach conclusions based on marginal data sets, the pressures to conform to agency guidelines—all can be negative forces operating on individual scientists.

An interesting sociological study, not yet accomplished, to my knowledge, would be an examination of postexposure careers of those scientists considered guilty of fraud. Anecdotal information from a number of cases indicates that

- Most persist in denying personal guilt, insisting that it was the action of a junior colleague, or assistant, or graduate student that was fraudulent; some even claim a conspiracy among assistants.
- Some are dismissed from current positions, but reappear in research and/or teaching positions with low visibility elsewhere.
- Some resign under pressure.
- Some disappear into nonscience occupations.
- Some disappear.
- Some retreat to nonresearch aspects of their specialties (MDs, for example, go into private practice, Ph.D. chemists move to industrial production, or Ph.D. psychologists become personnel directors or counselors).
- Some, surprisingly enough, continue in positions of authority, or are even promoted to positions of higher authority.

Disclosure or suspicion of fraud usually calls into being a review process which can be (and probably should be if necessary) long and detailed:

- The university administration and often colleagues in high places in granting organizations, after an initial reaction of disbelief, may take heroic steps to cover up the misdeed, or may delay any action interminably—especially if the accused is a member of the scientific elite.

- If some action is required, an internal committee may be formed to see if "probable cause" exists, but its report may be tabled for years. In time the university may conduct a formal inquiry, or it may not.
- A funding agency may request an investigation of charges, naming outside experts to participate.
- The accused may be reprimanded and suspended for a stated period, but not dismissed.
- The accused may continue in his or her position, and the complainant may be dismissed or transferred.
- The accused may institute litigation which can drag on for years, based on claims of unjust dismissal or damage to professional reputation.

Too often, the participants (the complainants) in disclosures of fraud are those least able to cope with "the system"—graduate students or junior faculty members. What is the proper response of juniors when confronted by or enmeshed innocently in instances of marginally ethical practices—by their supervisors or other investigators—the "augmenting" of data, the unwarranted unsurpation of senior authorship, the deliberate undermining of another's scientific reputation? Great circumspection and caution are watchwords. The "system" is highly conservative and resistant to perturbation, even when abuses seem evident. Any decision to expose abuses should be made only after great introspection and careful consideration of consequences. Rewards for damaging disclosures or "whistle-blowing" are scant; reaction usually seriously injures the dissenter; and the system closes ranks against and punishes those who attempt to destroy a comfortable equilibrium. Also, public disclosure of abuses—however well justified—immediately places the discloser among the ranks of the malcontents, the petty complainers, the embittered ones, the perverted activists, with real or imagined injuries from the system. These are people who run to authorities (deans, department chairpersons, legislators, or worse, to representatives of news media who thrive on controversy of any kind), with outcries against minor abuses of a system that is basically honest and remarkably self-correcting. Some professionals feel, however, that present-day science is stressed to the point where quality-control mechanisms no longer work, and the system is no longer self-correcting, except when abuses are pointed out by outsiders.

It seems generally true that major disclosures of fraud are usually consequences of minor events, and are made by wronged, disgruntled, idealistic, or publicity-seeking junior colleagues, administrative assistants, laboratory assistants, graduate students, or secretaries. Disclosures are rarely the result of peer examination or of the principal self-policing mechanisms of science—evaluation of research proposals by colleagues, review of methods and findings by the unit director, refereeing by peers of manuscripts submitted for publication in journals, and attempts by others to verify findings by repeating critical experiments. Often the scientific stature of an investigator can shield him or her from close scrutiny, by shortcutting the control devices of peer review and refereeing.

A classic and often-cited example of failure of the control systems of science to guard against abuses by the elite is that of the microbiologist Hideyo Noguchi (real name) in the early decades of this century. An associate of Simon Flexner at the then Rockefeller Institute for Medical Research, Noguchi spent an entire career of almost 30 years producing a flood of what are now considered mostly spurious reports of iso-

*lation and cultivation of principal human pathogens, includ-
ing those causing polio, syphilis, and yellow fever. For whatever reason—personal eminence, eminence of his institute,
lack of confirmatory work by others—he enjoyed a long and
remarkably successful career (by some standards) and died
before challenges to his findings arose. A recent analysis by
H. H. Smith in the journal* Science *titled "A Microbiologist
Once Famous"[6] disclosed little of value in Noguchi's work,
except a clear object lesson to science that the published
results of no one should be immune from scrutiny by
colleagues.*

Fortunately, too, some established scientists refuse to condone significant ethical abuses committed by colleagues. Cog-
nizant of the risks, this small percentage of the community will
still speak out in professional forums, when it is important to
do so.

All this talk of fraud in science can easily lead a reader to
wonder about the supposed veracity and objectivity of scientists as an occupational group. Emphasis on the pathologies
can inspire questions such as, "Are they all a bunch of liars and
manipulators?" At about this point it may be time to insert a
reminder that the anomalous events discussed in this chapter
are just that—anomalies—and they do not characterize the
standard professional activities of the great majority of practitioners. Discussion of this sensitive topic with many excellent
scientists, during the admittedly casual research on which this
book is based, has helped to reaffirm the concept that most
excellent scientists still consider honesty to be an absolute
unvarying requirement of the job, and that most would tolerate no deviations. All admitted to pressures for success and
achievement, but none condoned falsification in any form as a
way to achieve these goals. To those scientists questioned,
proven frauds represent deviant actions by a tiny minority of
scientists; similar types are of course present in all professions.

Moreover, some of those questioned pointed out that fraud could bring disgrace upon the parent organization and on the institution of science itself. Success, to most respondents, is important, but only if achieved through legitimate means— original ideas, productivity, syntheses—and not through anything even faintly unethical. Fraud debases the long tradition of science and the results of legitimate scientific research.

CONCLUSIONS

Controversies have occurred in every decade in the history of science. Some filter into the news media, but many more are confined properly to scientific forums. The causes of conflicts occupy a complete spectrum from minor disagreements about interpretations of data to major career-threatening accusations of theft or fraud. Two principal contributors to the development of controversies in science are failure of adequate communication and an imprecise definition of ethical perimeters. The resolutions of conflicts are not always crisp and definitive; the consistent loser is the image of science as an impersonal search for truth.

All controversies in science do not, of course, involve accusations of fraud, but the more serious cases do. Fraud exists in science, as it does in any activity which humans undertake, but in its severe manifestation it is anomalous and rare. Motivation for fraudulent behavior is complex, but in recent decades the principal factor seems to be stresses to be recognized professionally and to acquire (or continue) research grants. Some observers also point to the decreased effectiveness of the self-policing mechanisms of science as another factor.

The reality seems to be that controversies are inevitable in a less-than-perfect world, and are a part of scientific existence, but that proven instances of fraud, though rare, can never be taken lightly or condoned by a profession dedicated to truth.

REFERENCES

1. Robert K. Merton, Priorities in scientific discovery, *American Sociological Review* 22 (1957), pp. 635–659; Singletons and multiples in scientific discovery, *Proceedings of the American Philosophical Society* 105 (1961), pp. 470–486; Resistance to the systematic study of multiple discoveries in science, *European Journal of Sociology* 1963 (4), pp. 237–282; Behavior patterns of scientists, *American Scientist* 57 (1969), pp. 1–23.
2. William Broad and Nicholas Wade, *Betrayers of the Truth* (New York: Simon & Schuster, 1982).
3. Stephen H. Schneider and Lynne Morton, *The Primordial Bond* (New York: Plenum Press, 1981).
4. Robert K. Merton, The ambivalence of scientists, *Bulletin of the Johns Hopkins Hospital* 112 (1963), pp. 77–97.
5. Nelson Marshall, Advisor, advocate and adversary, *Bulletin of the Ecological Society of America* 54 (1973), pp. 6–7.
6. H. H. Smith, A microbiologist once famous, *Science* 212 (1981), pp. 434–435.

CHAPTER 10

THE ALSO-RANS
Burnout, Fade-out, Guaranteed Losers

Burnout: A Nonproblem for Most Scientists; Fade-Out: A Real Problem for Many Scientists; The Many Faces of Marginal Scientists: Case Studies in Waste Accumulation

INTRODUCTION

Frauds and serious controversies in science, discussed in the previous chapter, can destroy careers, or can quickly transform what seems like success into failure. Beyond these career-damaging crises, however, there are other categories of events or attitudes which result in mediocrity and failure. Prominent among them are burnout, fade-out, and other aberrant behavior patterns. This chapter is a brief guided tour through some of these shadowy back alleys of science—all in the interest of a balanced presentation.

BURNOUT

The concept of "burnout"—defined by psychologists in their own unique jargon as "depersonalization resulting from inadequate rewards or recognition for accomplishments, or as a consequence of continued striving toward unrealistic goals"—has been popularized in recent years. Burnout does occur among scientists, but its prevalence is lower than among professionals in most other occuptions. A partial explanation may be that in the reward system of science, satisfactions are only partially derived from the organization which pays the salary. A significant part of the recognition of worth and productivity comes from the larger scientific community, and a significant part of the rewards for achievement are internal, in the mind of the individual scientists. Thus, the predominant cause of burnout in industrial organizations—inadequate reward and recognition for outstanding work—is lessened in its potential for harm in scientific organizations.

The unique reward system of science is of course limited to the active scientists. The scientist-administrator or the scientist-bureaucrat at any level, without direct firsthand participation in science to derive internal rewards, is just as vulnerable to burnout as are his industrial colleagues, either as a consequence of inadequate recognition of accomplishments or of unrealistically ambitious career goals. Incipient burnouts can be seen in government agencies in particular. A junior agency staff member, recently recruited from a field office and full of idealism and enthusiasm, may be given a piece of a project to develop, only to find later that the real thrust of the project is elsewhere and his good work will not even be incorporated in the final draft. The midlevel bureaucrat might prepare an excellent planning document, policy statement, or position paper, and see it vanish "upstairs" forever, or see it reappear in modified form credited to someone else in the hierarchy. Even a senior career bureaucrat can see months of painstaking study and drafting discounted in seconds by a spur-of-the-moment decision of a politician or political appointee to kill the document. The playing out of these events in sequence has a snowballing effect, and provides the basis for eventual burnout.

The survivors that I've seen of this destructive process are those who can erect barriers; those who maintain some personal distance from their work; those who can accept anonymity as a condition of employment; and those who can tolerate rejection of good work for trivial reasons. Those who can't develop such blocking techniques are candidates for burnout, unless they leave for other kinds of jobs. Those who can develop appropriate blocks, unfortunately, are *not* those who will succeed; success and recognition are relatively unimportant to them, so they are well suited for the dead-end jobs.

Among active scientists, the only ones clearly identifiable as especially prone to burnout are the "scientific workaholics." They are difficult to generalize about and even more difficult to make judgments about, but they tend to be professionals who have substituted work and work-related activities for almost every other form of pleasure and preoccupation:

- Scientists who prefer to be in the laboratory in the evening to being at home, and, if at home, are elbow-deep in journals or draft manuscripts
- Scientists who at a social gathering have only one topic for discussion—research and research results (preferably their own)
- Scientists who may be judged socially inept or uncaring by less-productive colleagues

These often driven people should not be pitied or disparaged, since from their numbers can come some of the best products—research results, writing, conceptual thinking. Their pleasures in a total preoccupation with science are no doubt real, even if their values may seem skewed when compared with the social values of others.

A major finding of the present study must be that a *significant number of excellent scientists are borderline or in some instances admitted workaholics.* Some don't realize it; some realize it and apologize, as if it were a failing; some realize it and consider it a compliment to be so labeled. People in the last category often wonder how good science can be done in any way other than with total dedication. They tend to be secretly suspicious of colleagues with normal family lives, hobbies, and outside social interests. They tend also to be susceptible to burnout if expected rewards—especially the respect and approbation of colleagues—do not materalize.

Although difficult to demonstrate statistically, it does seem that the frequency of burnout among postdocs and junior untenured university faculty has shown an increase during the past decade. Assessment criteria for burnout are inadequate, but they can include such elements as

- Marked and sudden decrease in interest in teaching or research
- Last-minute request for leave of absence
- Precipitous resignation just before or during the academic year
- Bizarre behavior in the classroom or laboratory

These actions may be symptomatic of increased stress in acquiring and holding good positions, and the unhappy combination of high expectations with low rewards from organizational administrators.

FADE-OUT

A syndrome less glamorous than burnout, but not less important in its impact on productivity, is one which can be labeled "fade-out," characterized by gradual withdrawal from active competition. Its victims are often of middle age, although symptoms may be detected much earlier.

The typical fade-out sees himself or herself as falling back and disappearing among the pack in scientific activities, doing research which has become routine, writing papers which are mediocre, discerning coolness rather than acceptance from colleagues, and finding scant satisfaction in any professional pursuits. More and more time becomes occupied by "outside" interests—hobbies, television, gossiping, daydreaming—and scientific accomplishments dwindle and disappear.

Fade-out is an affliction of the average or the mediocre scientist, whereas burnout is usually found among the strivers. The fade-out-prone individual tries for a time, often in a halfhearted way, and then decides that science is not worth the effort. From then on, commitment is minimal, productivity slumps to barely acceptable levels, and professional contacts diminish. Such a person rarely displays enthusiasm, is quick to point out deficiencies in facilities or management, and becomes a weak link in team research. He or she can be found jogging, playing tennis, or gardening, but will rarely be visible at seminars, committee meetings, staff conferences, or the laboratory bench.

Causes of fade-out are complex. Certainly a feeling of inadequacy must be a partial cause, as must deficiency in background, or energy, or intelligence, or interest. Such feelings may be exacerbated by association with bright, energetic, frequently younger colleagues, who seem to publish profusely and who move upward in the system rapidly. Further exacerbation can result from failure to achieve honors or salary increases. A self-reinforcing downward spiral is the normal consequence.

Part of the problem is undoubtedly flawed perceptions of what is productive and what is marginal in science. Fade-outs are often convinced that their productivity has been maintained at more than adequate levels, and that their contributions continue to be equivalent to those of their peers. Rationalizations and self-deceptions are easy in an occupation where job performance standards cannot be rigid and precise.

A state of minimum productivity and commitment can persist for long periods. Scientific groups—academic, government, or industry—are liberally populated with fade-outs. Some are tolerated, some are hidden, and some are given menial tasks. Once in a while some are fired too, especially if they are in industrial research groups. The usual practice in government or in universities, however, is to suffer the continued presence of fade-outs, hoping (usually vainly) for rejuvenation.

A special kind of fade-out makes its appearance soon after completion of the doctoral dissertation (after a normal post-stress period of rest and recuperation). The thesis represents *the* research peak for a surprising number of individuals, and everything subsequent to it is downhill. Some post-Ph.D. fade-outs remain in research but never publish a significant research paper, regardless of pressures; others escape quickly or gradually into trivial research-related bureaucratic activities; and still others disappear into minor supervisory or managerial roles without ever publishing their thesis material. It is difficult for a nonpsychologist to understand the motivations (or lack thereof) behind this phenomenon; it is easy for any scientist to deplore the waste in such a circumstance. It must be, in these cases, that the degree is considered as a form of job insurance or an entry permit, unaccompanied by any commitment to further efforts in science.

GUARANTEED LOSERS

The "fade-out syndrome" is only one facet of mediocrity and failure in science. Many other categories exist, and brief attention to the "guaranteed losers" may be justified if it leads to useful insights about the etiology of mediocrity or failure. Although the case history method used as a basis for this book has concentrated on successful scientists, enough data have been accumulated about those who are not so successful to provide some understanding of the characteristics of "losers." It is even possible to categorize these "problem scientists"— the "guaranteed losers," the "downhill racers"—as "the dilettante," "the hobbyist," "the unwary activist," "the dropout," "the chronic underachiever," "the hollow strategist," "the unsuccessful manipulator," "the itinerant instructor," "the perennial postdoc," "the PR person," and "the whistle-blower." The trivial or marginal scientists in these categories can be fascinating studies, but analyses of their careers can be depressing too, because of the obvious waste of talent.

- *The Dilettante.* Master of the superficial—flitting from one short-term project to another—the dilettante rarely makes a significant contribution in any research area, but has a passing interest in many. He or she often has the self-image of an innovative thinker, with too many ideas to explore any of them in detail, and too many research interests with too little time to develop substantive data about any of them. The dilettante enjoys discussions about broad concepts and generalizations, but becomes restless when details, work schedules, procedures, and writing assignments are topics of conversation.
- *The Hobbyist.* Most occupations are generously populated with people who consider a job as a means to an end—a way to earn a living without personal commitment. Amazingly, a number of scientists look upon the practice of science in just that way, instead of seeing it as the exciting, satisfying, all-consuming vocation that it

really is. If the doing of science is just another 40 hour a week job, to which no further loyalty is felt, then any diversion is welcome, and energies are conserved for the important things in life—home repair, golf, jogging, sex, skiing, tennis. This may be a legitimate approach to living, but it does not lead to success as a scientist.

- *The Unwary Activist.* The number of causes which are available in today's society is limitless—from abortion to nuclear proliferation to environmental degradation to saving whales to human overpopulation to women's rights to gay rights and beyond. Spokespeople for such causes are delighted to find new recruits, especially those who are vocal and who bring to the movement some kind of special background. Scientists, particularly very junior or very senior ones, often become enmeshed beyond their expertise, to the detriment of their scientific credibility. In some instances, scientists cum zealots are actually used for the purposes of the movement, regardless of costs to their professional reputations. (It could be argued, however, that some of those with scientific backgrounds who *knowingly* metamorphose into extremists can thrive in new careers, more correctly in the area of public relations rather than science.)

- *The Dropout.* Clustered on the fringes of science are numbers of people who for one reason or another have not completed professional training and who exist unhappily as technicians, part-time volunteers, or perennial junior staff members. They tend to be perpetually dissatisfied with their roles, but are unwilling to dissociate themselves totally from the scientific community. Some do drift away and disappear; others persist in marginal jobs; a few even make decisions to complete graduate work. The norm, though, seems to be to spend an entire career in the backwaters, doing the "scut work," becoming unhappier each year with their lot, and wondering what went wrong.

- *The Chronic Underachiever.* A little more difficult to describe is the professional who just never seems to put forth the added effort to make an average piece of work an excellent one. For whatever reason—lack of time, talent, or interest—research papers are slow in appearing and mediocre in quality, lectures are routine and uninspiring, contacts with peers are sporadic and desultory, and participation in scientific community affairs is minimal.

  Sometimes this approach to science seems to be a continuation of behavior patterns developed in secondary schools, where no more than marginal application is expected. Possibly this is even true at the college undergraduate level, but it is difficult to understand how such people are accepted by graduate schools, especially in the recent selective climate, and, more importantly, how they actually complete graduate training. At any rate, they have penetrated the screen and there they are, a large pool of the "persistently mediocre," putting in their time and following the curvature of the earth. Daily contacts with such underachievers lead always to a sense of waste—of opportunities not taken and of joys not experienced.

- *The Hollow Strategist.* Research is a tough, grinding, time-devouring, ultimately satisfying occupation. Most scientists accept this. Some, though, are not willing to make the total commitment required, and spend what should be productive time in devising strategies to avoid doing research. Some become totally immersed in faculty committee affairs, and complain that they have no time for the laboratory. Others become overly involved in personal problems of their students, and again have no time left for the laboratory. Still others carry on extensive almost nonstop discussions with colleagues—in offices, corridors, or cafeteria—and find that their research day is remarkably foreshortened. Any planning done is oriented toward methods to reduce actual research time, rather than to enhance it.

• *The Unsuccessful Manipulator.* Scientists tend to be somewhat naive insofar as interpersonal strategies are concerned. They are normally open and honest in their relations with colleagues, and expect reciprocation. Occasionally, one of these colleagues will depart from expected practices and try to use others for selfish purposes, usually to gain some form of scientific aggrandizement. The manipulation may succeed, but only once, since colleagues will be forever wary after that episode.

The unsuccessful manipulator may move on to other susceptibles, but the route is downhill as more and more colleagues become aware of the device. Common practices are proposals for joint research, in which the manipulator supplies an idea but none of the work; or preparation of a book, for which the manipulator writes only the introduction, but gets his or her name on the cover.

• *The Itinerant Instructor.* Today's academic environment is grim. Students are increasingly scarce, tenure is almost nonexistent, and salaries are miserable. Faculty "positions" are more and more short-term and in a class known as "nontenure track." Good universities are still turning out good Ph.D. recipients, but they are finding a marginal to poor employment scene. Some applicants retreat to the temporary security of a postdoctoral appointment; others accept short-term instructorships, which lead only to other short-term instructorships. To many well-qualified journeyman scientist-teachers this becomes an endless treadmill, of necessity but not of their choice. They move from university to university, rarely finding a home, and becoming more and more uneasy about the future.

Fortunately the picture is not totally in shades of gray. Here and there among the wanderers are success stories—of excellent people who manage to get off the transient treadmill and find reasonable to good jobs with some prospect for advancement and security in the profession for which they were trained.

One of my best recent examples is Dr. Frank Anselmo, recent Ph.D. from a good eastern university, with specialization in polymer chemistry, and with a career goal of teaching and research in an academic environment. His immediate post-doctoral years were rough. He spent one year as a postdoctoral fellow at a midwestern university , then nine underpaid months as an instructor at a small northeastern private college (replacing a faculty member on sabbatical), then a year as an instructor at a giant southeastern junior college.

Then one of those rare events happened; he was selected from an original field of almost 100 applicants for a tenure-track position as an assistant professor at a progressive eastern college—one which encouraged faculty research. He is now in his second year there; his courses are considered exceptional; and he has just received a small faculty grant for a research project.

I asked him recently for some explanation for his good fortune. His analysis listed (in priority) (1) luck, (2) refusal to quit and take an industrial job, even when things were tough, (3) self-hypnosis to maintain a positive outlook on life, and (4) persistence in continuing some minimal research activity, even when facilities were abominable.

The story is not too different for other excellent junior people who have escaped from the transient trap. Luck and chance events characterize many stories, but persistence, a positive attitude, and some continuing if minimal research involvement are frequent features too.

- *The Perennial Postdoc.* A subset of the "itinerant instructor" is the "perennial postdoc," who moves from one research group to another, gaining useful expertise, but not acquiring a permanent berth. The phenomenon is particularly apparent in some of today's high-technology fast-breaking research areas, where many skilled hands are needed for countless experiments and endless replications, but where little hard funding and few permanent positions are available. Overlong existence in postdoctoral roles is sometimes encouraged (or at least condoned) by senior faculty members, who may be reluctant to see needed skills disappear. The matching piece of the puzzle here is the overly dependent postdoctoral assistant with good training who finds existence tolerable financially and stimulating scientifically, so is in no hurry to face life as an independent investigator.
- *The PR Person.* Many university faculties are blessed with the presence of a junior member who assumes, willingly, an array of functions considered tedious or semiprofessional by more senior members. Included are responsibilities for seminar schedules, travel arrangements for speakers and visiting lecturers, guided tours of facilities for visiting dignitaries, meetings with activist groups, cataloging and maintaining visual aid material, presentations at secondary school career days, and judging secondary or elementary school science fairs. Each added responsibility decreases the time available for pri-

mary functions—research and teaching—and each one lessens by some degree the scientific productivity which will be used as the principal determinant of upward mobility.

Some junior faculty members assume these extra-curricular responsibilities gracefully, recognizing a reasonable commitment to departmental well-being, but some are too insecure to object when the load becomes excessive. Some actually invite added responsibilities of this kind as an escape from research.

• *The Whistle-Blower.* Most scientists are honest, sincere, and dedicated—not given to murder, theft, or other major crimes. They are usually carefully ethical in the practice of science. Occasionally, in any scientific group, an individual will be found who can identify evil in any situation. He or she is frequently an active participant in the rumor network, and is quick to find malfeasance or worse in almost any actions of colleagues, especially the more successful ones. Injuries, slights, minor lapses in communication—all are ingredients carefully hoarded by such people, to be deployed gleefully in informal or formal grievances, or anonymous communications with whatever hierarchy seems receptive at the moment.

This type of petty activity is distinct from the disclosure of rare major frauds in science, which should be the joint responsibility of the entire community. The petty informer deals more with minor harassment, and is ready to view any event or activity with suspicion.

These, then, are some of the kinds of marginal scientists—the guaranteed losers—who are sources of frustration to their colleagues, and to administrators. Their numbers, unfortunately, are significant; they are not rare exceptions. Included in these dismal categories are bright and well-trained people—

but people who are denying themselves, or are being denied
by the system, day by day, access to the choice fruits of a
superb occupation.

CONCLUSIONS

The obligatory inspection tour of the "down side" of science, continued in this chapter, focuses on three identifiable
states: burnout, fade-out, and guaranteed losers. Burnout is not
common among active scientists, except for the workaholics,
but fade-out is too common, especially among mediocre scientists. Fascinating examples of other behavior patterns of
marginal scientists exist—including, but not limited to, those
of the chronic underachievers, dillettantes, dropouts, manipulators, and transients. These and other categories of "also-
rans" populate the outer fringes of what is for many scientists
a dynamic rewarding occupation.

One final word must be said for those forced into the
"also-ran" category by the present depressed state of academic
science. Many talented, well-trained, very good junior scientists find themselves classified as transients and temporary
occupants of nontenure-track faculty positions. To lump these
people with the long-term marginal scientists described in this
chapter is more than a little unfair, according to a most perceptive critic (my editor). I have no alibi, and I agree that they
deserve special consideration. A number of them will make it
despite the odds—those who are the most talented and the
most tenacious.

CHAPTER 11

THE AGING SCIENTIST AND THE SEARCH FOR IMMORTALITY

The Reaper at Arm's Length: Science Is Forever, If You Practice It; Alone on the Ice: A Nightmare Come True for Scientist-Administrators; A "Code of Practice" for Senior Scientists; The "General Withdrawal Syndrome"; The Search for Immortality

INTRODUCTION

It seems entirely proper that this part of the book devoted so doggedly to the darker side of science should wind down with an exploration of the last acts—the aging scientist and the futile but perennial search for immortality. I remind you that this is a book titled *The Joy of Science*. Does the joy persist to the final curtain, or do scientists suffer the traditional fate of elderly Eskimos of being left to die alone on the ice?

The concluding phases of any professional career should be of legitimate interest to all members of that profession, and science is certainly no exception. This book has attempted to scrutinize careers of successful scientists, and it seems especially important to give careful attention to the last act—how it succeeds or fails in integrating all that has gone before, and how it maintains or fails to maintain the momentum of earlier years. Among the many areas which should be explored are these: treatment of and attitudes toward older scientists by universities, government organizations, industry, and foundations; graceful versus awkward role changes with advancing age; expectations of outstanding senior scientists as contrasted with those who are merely very good; the closing era of the science administrator's career; and the legitimacy of a search for immortality.

THE AGING RESEARCH SCIENTIST

Advancing age is undoubtedly least threatening to the scientist who has resisted administrative responsibilities and who has been a consistent contributor to the substantive published literature. If original work has been augmented by reviews and books, and if participation in scientific societies has been active, the closing years of a career can be peaks rather than terminal moraines, as measured by stature in the field, awards, and recognition by colleagues.

The "retirement syndrome," so threatening to science administrators, is not a serious problem to a senior scientist, who is judged by peers on the basis of professional accomplishments and contributions. Sustained productivity, continued high competency in a subdiscipline, and research perspectives based on years of active participation, all provide strong defenses against administrators who might seek to reduce or eliminate positions or programs.

There can be, of course, overriding negative forces at work. Political decisions can affect government science and scientists drastically, regardless of the quality of research performed. Entire projects or laboratories can be eliminated and grant funding can be reduced or terminated at the whim of the politician-scientist appointed to head an agency. Excellent senior scientists can be affected by such actions, even though they tend to be less vulnerable than junior colleagues, because of tenure, pension plans, or ability to find other employment.

Older scientists must guard against declining productivity. There is some statistical evidence for reduced output with advancing age, with a peak in the thirties. Productivity in science is so much an individual matter, however, that statistical associations with age should not be overemphasized—especially when all factors influencing publication of scientific papers are considered.

Dr. Samuel H. Albertson is the kind of scientist who can best be described as indestructable. His record of significant publication now spans more than 30 years, without sign of diminution. Beginning with a published thesis considered even today as a basic reference, he explored over the following two decades many broader ramifications of his thesis topic (an area of biophysics until then virtually unknown). During that period he served successively as project manager, group leader, division chief, and laboratory director, always retaining a hands-on participation in active research, despite administrative pressures. Twelve years ago he voluntarily gave up the director's job and became a "principal scientist" in his organization. His productivity since then has increased significantly; he has published two books already considered classics, as well as several substantive review papers and almost 30 original scientific contributions. Although well past retirement age, he charges along with undiminished enthusiasm.

THE AGING PROFESSOR

Aging professors, still fully involved in teaching, may find student contacts to be less stimulating than they were two decades earlier. Despite some of the privileges of seniority (more graduate and fewer introductory courses; more seminar and fewer lecture courses; reduction in laboratory course commitments), some symptoms of disenchantment (and even boredom) with the academic routine may appear, including

- A progressive feeling of distance from students—in age, language, life-style, and life goals
- A disturbing feeling of distance from junior faculty members, in outlook and teaching methods
- A worry that lectures have become stories told too many times
- Rebellion against what seem to be increasing committee assignments
- A tendency to relegate larger and larger segments of lab and lecture courses to teaching fellows

This trend toward disenchantment is of course not universal. Exceptions are abundant. Among them are

- The really exceptional lecturer, able to hold the full attention of classes of any size
- The professor with an international reputation in his or her field, who speaks authoritatively and commands respect and admiration of graduate students
- The professor with active field or experimental research programs of the kind which can involve numbers of students
- The professor who has maintained an active interest in innovative teaching methods

- The professor who has a genuine interest in students as people, and who is able to offer personal advice on request

---

Older professors can be superb teachers and outstanding role models. One of my favorites, whom I have been privileged to observe for almost two decades, is Professor Harold H. Hawkins, now Distinguished Professor of Geology at a major western university. Dr. Hawkins has to be described as truly ageless. His relations with students are friendly and personal; his research always includes participation by an enthusiastic cadre of graduate and undergraduate assistants, regardless of the vagaries of grant support. His own enthusiasm for what might seem to be a dull subject—changes in coastal sediments related to weather—is infective, in the classroom or in field studies. No one in the department mentions retirement to him; he is too fully occupied with today's activities and with plans for tomorrow and next year.

THE AGING SCIENTIST-ADMINISTRATOR

Unlike the older research scientist or the older professor, older government/industry science administrators seem particularly vulnerable to diminution of career rewards and satisfactions with advancing age. Faced with increasing pressures to perform at maximum capacity, and to compete successfully with aggressive newcomers, some administrators elect to return to full-time research, or to accept noncompetitive advisory roles. Others withdraw completely and develop alternate careers (in teaching, for example). Still others elect to continue the race, and do so with variable success. Those administrators in the last category—the "fighters" and the "survivors"—can expect eventual confrontation with a "retirement syndrome" whose dominant sign is progressive "distancing." Elements of the syndrome for those who are over the hill (defined as being over 50) or are being edged out, include the following:

- Control over program budgets and staffing is reduced
- Advice on policy and administrative matters is ignored or rejected
- Travel to conferences and workshops is restricted
- Staff assignments are given to relatively junior staff members
- Critical positions are awarded to comparative newcomers or to assistants when vacancies occur
- Invitations to planning and policy conferences become scarce or nonexistent

If it is apparent that a change in role and authority is being clearly dictated by the system, the change should be effected gracefully. Government and industry are not noted for grace in these or in most other situations, so most of the grace will be the responsibility of the person being affected by the change. This may prove to be one of the most difficult responsibilities of a science administrator's career.

Dr. Nelson Bornholm, former member of the executive hierarchy at a large western university, and now professor of physics, exemplifies to me a smooth career transition from administration to teaching, near the end of a long and varied academic career. Earlier, after two decades of successful research and teaching, Professor Bornholm elected to accept a substantial but ill-defined role as "coordinator" of a multidepartmental institute—a matrix type of organization in which participants were funded by the institute but were still administratively responsive to their original departments. This was obviously a difficult (some would say impos-

sible) job which required the utmost in tact, pursuasiveness, and integrating abilities. He had many competencies which fitted him for the role—credibility in his own discipline, a frank, friendly approach to colleagues at any academic level, and an urbane manner which kept him at ease in any environment and in any group. During the decade of his tenure the institute prospered and expanded. Sensible policies were defined; additional departments and faculty members became participants, and the organization achieved national recognition.

Unfortunately, while gaining stature and competence as an administrator, Professor Bornholm was gradually withdrawing from affairs of his discipline, to the extent that he no longer felt comfortable about accepting graduate students or refereeing papers in his specialty. A moment came, three years ago, when he decided that despite administrative conquests, it was time to return to his true vocation—research and teaching in physics. After a sabbatical spent in getting up to speed again in his specialty, he returned to the classroom, secure in the conviction that this was a correct career decision.

A PROPOSED "CODE OF PRACTICE" FOR SENIOR SCIENTISTS (50 AND OVER)

For those productive scientists who reach that magic age when options begin to evaporate and the system begins to "take away our playthings one by one," adherence to a new set of operational principles—in fact, the declaration of a "senior scientist manifesto"—seems to be required. Some proposed elements of the manifesto could be

- Accept the reality that yesterday's successes and awards are only slightly more valuable than yesterday's newspaper.
- Make every age one of personal growth, metamorphosis, and liberation from the ordinary.
- Avoid anachronistic thinking; do almost nothing the way you did it five years ago.
- Try to remember that the best approach to most situations is a positive one.
- Never catch yourself saying "the way we used to do it" or anything resembling that phrase.
- Believe that personal scientific goals, some set long ago, are still attainable.
- Give advice if you're asked, but avoid staff positions with "advisory" in the title. Such jobs provide quick access to the shelf.
- Maintain some form of continuing meaningful scientific dialogue with the newest and brightest members of the group.
- Resist, with some grace but with vigor, arbitrary demotions or loss of "perks" because of age alone.

THE MESSIAH COMPLEX

Older scientists, whether retired or not, often take up causes which are related directly or indirectly to their professional expertise. They may become protagonists or antagonists in current issues of society—population control, world nutrition, nuclear disarmament, abortion—trading on reputations and credibility acquired during active scientific careers. Involvement is usually based on strong convictions and on perceptions or perspectives derived from professional knowledge. As such, senior scientists can be effective spokespeople in matters related to their areas of competence. Their opinions and analyses are highly regarded by the concerned public.

With the deference of less-informed people, however, comes a commitment to speak responsibly as a professional, not to espouse extreme positions on issues, to avoid statements which cannot be supported by the best available data, to avoid dogmatism, and to resist at all costs any metamorphosis into self-proclaimed prophets or messiahs.

Temptations are great to exceed perimeters outlined by the evidence at hand when discussing social issues with high emotional content. Convictions, based on personal opinions, can outweigh good judgment, and lead to embarrassment. At some point, scientists may cease speaking as professionals and become nonobjective moulders of public opinion. Some would see this as a perfectly logical evolution, while others might view it as a prostitution of science, in the sense that laurels acquired in a technical field are traded as devices to influence decisions on social issues.

THE "GENERAL WITHDRAWAL SYNDROME"

A special and rarely discussed subset of the "fade-out" principle described in the previous chapter can be diagnosed with surprising frequency in aging scientists (defined by today's youth-oriented culture as anyone over 50). Best identified as the "General Withdrawal Syndrome," it is a concept comparable in breadth to Hans Selye's stress-provoked "General Adaptation Syndrome"[1] but with a larger psychological component. It is clearly a consequence of accumulated slights, kidney punches, and negative job-related decisions based on advancing age. This kind of fade-out can affect previously excellent scientists as well as mediocre ones. Often it precedes a decision to retire or change occupations.

Onset of age-associated withdrawal is gradual; its origins may be difficult to trace in individual cases, but they begin with a first reference by a department head, colleague, assistant, or student to some real or assumed inadequacy related to advancing age. From that moment the steepness of decline seems dependent on the frequency of similar events, real or inferred, and the degree to which the recipient is destroyed internally by them. Some classic one-liners are

- "Let's leave the seminar schedule to younger heads and shoulders."
- "How will the department benefit from supporting your specialized training at the present stage in your career?"
- "Science must have been so uncomplicated when you were in graduate school."
- "My mother took your course 20 years ago."
- "I had no idea that you were a backpacker (skier, long-distance runner, touch football player)."
- "We're going to fill that key program slot with a fresh new Ph.D. and not with a member of the present staff."
- "We'd better get the project under way, you are a 'short-timer' and won't be around forever."

Some of the characteristic signs of the general withdrawal syndrome include gradual disappearance from professional meetings, retreat to trivial research projects, increasing preoccupation with sedentary hobbies, and reduced communication with colleagues. The etiology can be varied. The normal age-related decline in energy contributes, as does awareness of and difficulty in coping with the explosive rate of expansion of information in any discipline. Some of the causes can be intensely personal—sensed exclusion from "invitation-only" symposia or workshops, the appearance of a 39-year-old dean or laboratory chief, the rejection of a manuscript by a journal which has published many previous papers, the realization that a long-projected classic technical book will never be written, or harsh treatment of a research grant proposal by a review committee of junior colleagues.

Factors contributing to the general withdrawal syndrome may include one or many small, individually insignificant events—tiny regressions which aggregate to the critical mass which precipitates a decision to resign, retire, or otherwise disappear from science. Some of the small signposts are the vague horrifying realizations that

- Lectures are not as sparkling as they once were.
- Physical features betray advancing age.
- Thought processes seem slower and fuzzier.
- Decisions (like whether to walk right or left at the corner) are harder to reach.
- On occasion, hands or voice tremble.
- It is increasingly difficult to stay awake in staff meetings.
- Colleagues seem to be exhibiting undue deference.
- Scientific meetings are less fun and are populated by much younger people.
- Pontificating, frequently and gratuitously, has become a habit.
- Career advancement is going to others, and age seems to be a determining factor.
- Energy reserves seem to be lower than they were formerly.
- Minor infirmities and random aches reduce physical activity and make some days seem very long.
- The last promotion was a decade ago.
- The tiny assigned faculty office has been occupied for one too many years.
- The new and younger department chairperson/supervisor used a tone that was actually *patronizing*.
- The listing in the college catalog of faculty degrees and the years of their awards contains what is clearly a high

percentage of "recent" people and very few "older" ones (your contemporaries).

The threshold for awareness of and responsiveness to these aggregated events varies with the individual. Some scientists can be stoic and unmoved; others can be destroyed by a sudden insight about the deteriorating state of their professional existence. The realization that nothing is forever, even in science, can be hard to accept, especially near the end of a career which has been characterized by achievement and success.

THE SEARCH FOR IMMORTALITY

Scientists are generally in accord with most people in the desire to leave behind, after they die, some indication that they existed—some small legacy, some tiny monument, some mark of tangible contribution to progress in understanding the universe. Usually, though, this is a vain hope. Names of discoverers of new principles, laws, theorems, or hypotheses are immortalized, but few in any generation are so honored. Biologists may be immortalized by having their names appended, as the authority, to a genus and species if they describe or redescribe one. Rarely, a taxonomist may name a new genus or species after an individual, thereby conferring a form of instant immortality on that person. [The author's dubious but best claim to immortality—at the moment—is that a taxonomist friend named a pathogenic marine amoeba in his honor, with his family name as the species name (*Acanthamoeba sindermanni*). But even this kind of miniscule claim to immortality is fickle and unstable. The next amoeba taxonomist on the scene may decide that *A. sindermanni* is identical to a previously described species, and suppress *A. sindermanni*. It will then disappear forever, even before its human namesake! *"Sic transit gloria scientia."*]

Some scientists think that their published work will confer some degree of immortality. Nothing could be further from the truth! Scientific publications are notoriously transient, with an average half-life of roughly seven years, and with the readership decay rate that is geometric (as measured by citations in current literature). A singular observation, confirmed by some librarians, is that *many articles are never read even once by any scientist*, though their author abstracts may appear in numerous abstracting publications months or years later.

A cursory review of literature-cited sections in recent volumes of several national journals would suggest strongly that science as we know it was created by an act of spontaneous generation about 1970 (supporting the seven-year half-life concept). In reality, some of the best work in many subdisciplines was done before most of the present-day practitioners were even conceived—but citations of this work are infrequent enough to form the extreme asymptotic tail of any readership curve. How, then, can immortality be achieved through scientific paper publication?

Survival time for books is a little better. Individually authored books provide the best choice for the serious seeker of scientific immortality. The half-life of a good technical book can be estimated at ten years—almost but not quite double that of an original scientific paper, but still a transient and minute blip on any scientific progress chart. A book normally summarizes the status of knowledge in a given speciality or subject matter area. Its life span depends to some extent on the rate of development of new information, and on the possible appearance in a few years of a new (and possibly better) book covering the same specialty or subject matter area.

Immortality, or at least its temporary equivalent—some tiny but lasting contribution to understanding the universe—

is probably an illusion for most of us. If we look dispassionately, especially near the end of a career in science, at our lasting contributions, they often seem insignificant or even nonexistent. A brave little pile of reprints, some already yellowing;
a book or two, already obsolescent on the date of publication;
an occasional medal or prize or life membership in a society,
quickly forgotten by colleagues; a few good graduate students
who rarely look back—these are our legacies. Only for a chosen few in any generation is there likelihood that the next generation will remember, or use, or refer to our specific
contributions.

If the search for immortality is such a fruitless activity,
what then are reasonable goals for a scientific career? Certainly
a contribution to the general fund of human knowledge is a
legitimate goal, as is acquisition of inner satisfactions for meaningful work well done. Although a little corny in its expression,
the joy of standing at the frontier of human understanding, in
at least some small area of science, must also be listed as a goal,
along with the internal pleasure of insights—of seeing solutions to problems, however tenuous they may be. Beyond
these, there are quiet rewards in participating in the development of a future generation of problem-solvers—the graduate
students, postdoctoral fellows, research assistants, and junior
staff members—who quickly absorb what is known and then
move with enthusiasm to extend its perimeter.

CONCLUSIONS

Aging, for some scientists, can be a personal battle against
disrespect, rejection, and oblivion, recognizing the frailty of
yesterday's successes and praises but remembering the personal importance of those successes. The psychologist Ernest
Becker expressed it well, saying "What man really fears is not

so much extinction, but extinction with insignificance. . . ."
The hope persists among professionals that their science will
outlive them, that their ideas, their publications, and their students will have some lasting importance.

Robert Merton, in his book *On the Shoulders of Giants*,[2]
described the contributions of the great conceptual thinkers to
progress in science, expanding on the metaphor that the future
landscape is best seen by standing on the shoulders of giants.
It seems equally tenable that progress in science can be made
and is made by standing on the shoulders of people of normal
stature, if accompanied by insight and productivity—and if the
debt owed to those predecessors is sincerely acknowledged.
This is the immortality that successful scientists achieve—not
a personal one, but as a functional part of an onward-flowing
activity of the human species.

The route, then, is not all downhill to mere oblivion, and
is not as grim as might be portrayed. Older scientists make
unique contributions—rationality, perception, insights, experience, ethical positions, a sense of history—important to science in any age. Some continue to practice good science for the
rest of their lives. Their role is not insignificant, and not one to
be minimized.

REFERENCES

1. Hans Selye, Stress and the general adaptation syndrome, *British Medical
   Journal* 1 (1950), pp. 1383–1395.
2. Robert K. Merton, *On the Shoulders of Giants* (New York: Free Press, 1965).

PART FOUR

ENCOUNTERS WITH THE EXTERNAL ENVIRONMENT

The third section of this book, "Descent from Career Peaks," slogged almost interminably through some of the backwaters of science, in an attempt to achieve some perspective on excellence and success. After this time in purgatory it seems appropriate, especially in a book on the joy of science, to get back on an upbeat course, by considering some important, stimulating, and occasionally pleasurable encounters with the world outside the perimeters of the laboratory and classroom.

For the professional who can and will, the larger universe includes involvement in science-related social issues of the time, in the political processes which impinge on science, in legal actions which may affect scientists, and in attempts to translate science into popular language. Each of these activities on the fringes of science requires extensive on-the-job training plus natural abilities; each requires great credibility and flexibility; and each can bring additional satisfactions to careers in which satisfactions are the norm.

These "outreach" activities can become absorbing avocations for those who are already successful as scientists and who want more interaction with social and political factors that help control their disciplines and can affect the future of science.

CHAPTER 12

SCIENTISTS OUTSIDE THE LABORATORY

Emergence from the Ivory Tower: The Advocate Role in Science-Related Social Issues; The Expert Witness as an Effective Participant in Legal Events; The Accommodations of Science to the Political Process; The Interpretation of Scientific Information for Public Use

INTRODUCTION

We are at the point where all the lines of inquiry developed thus far in this book should begin to converge, where we should begin to see a larger landscape in which those who have found success in the technical aspects of science go on to additional achievements within their fields, or begin to explore broader science-related activities. Examining the careers of successful professionals in the numerous categories described in Chapter 6 on "Destinations," a major dichotomy can be seen, between those who continue to invest in interests directly related to science, and those who move beyond and outside, to advisory or leadership roles in the larger human community.

Those who choose the broader course—on the periphery of science—deserve some attention in these closing pages. Thumbing through the masses of lightly processed data which form the fragile foundation of this book, it is possible to discern points where excellent scientists have moved far beyond the laboratory and lecture hall—into areas of social concerns, political interactions, and public interests. They have done this in part because of a perceived need to establish the relevance of their work, and that of colleagues, to a society searching for a factual base for decisions about issues of the moment. Those who persist do so because they find that they can perform effectively in these new roles.

The critical observation here is that *many excellent professionals seek out and participate vigorously in science-based sociopolitical activities; they derive pleasure and satisfaction from this added dimension of careers in which they have already achieved credibility and success.* Examples of larger social roles for scientists, that are based on scientific competence and credibility, but extend beyond disciplines, include (but are certainly not limited to)

- *The Advocate*—interpreting scientific data relevant to social issues
- *The Expert Witness*—providing scientific information, advice, and opinions to formal bodies (courts, legislative or agency committees)
- *The Political Liaison*—relating science affairs to political processes
- *The Public Scientist*—translating, simplifying, and interpreting scientific information for the public

Participation in these and similar "outreach" activities demands excellence, but it can reveal unexpected horizons to those who are qualified and who find that they enjoy the new perspectives and the climate outside the laboratory.

THE ADVOCATE

Scientific data are of course objective and factual—numbers, observations, descriptions, lists, mechanically-derived graphs, etc.—and as such are not transmutable. Flexibility and subjectivity enter scientific discussions when *conclusions* are drawn from the data and *interpretations* are made of the meaning of the data. Bitter and prolonged arguments develop regularly in scientific publications, seminars, symposia, and conferences about interpretations of data. These arguments and debates often lead to further data collection and then reexamination of earlier conclusions. Eventually (but not always), some common agreement is reached about the meaning of the data.

It is not unusual, however, in public issues which have a scientific foundation, to find reputable scientists on both sides of those issues, arguing from a common data base (or selected parts of that data base), reaching conclusions that are drastically different. Listening to such arguments, it is easy to sense a measure of *advocacy*, especially when conclusions are based on statistical correlations or extrapolations from limited samples. At some point in the interpretive process some scientists begin to move away from total objectivity and begin to fashion arguments for or against issues—arguments which depend on interpretations of data favoring personally held views. At that point those scientists become *advocates*—partisans and proponents (or opponents) instead of impartial experts. The process can be gradual; sometimes the metamorphosis takes years to complete, and at times it may not even be obvious to the ecdysist. If held within reasonable bounds, advocacy positions on important issues can have salutary effects on public debate and can make useful contributions to understanding of those issues.

The advocate occupies a position slightly to the left of center in a spectrum which ranges (from right to left) from the "noninvolved professional" through the "concerned professional" to the "advocate" and finally on the far left the "extremist." The *noninvolved professional* refuses to take stands on issues, preferring to hover almost invisibly behind data sets offered for eventual interpretation by others. The *concerned professional* makes occasional limited forays from secure data bases to offer interpretations and recommendations. The *advocate* makes frequent charges from adequate data bases; considers interpretations and recommendations to be part of any scientist's responsibility; and may offer carefully phrased opinions, predictions, and speculations in public forums. The *extremist* has usually lost objectivity on a particular issue; has become an obvious protagonist or antagonist; and may be denied, increasingly, space on scientific agendas.

By far the largest proportion of excellent scientists interviewed to provide background for this book—when asked to locate themselves somewhere on the continuum of degrees of involvement in public issues—classified themselves as "concerned professionals." The term itself has positive connotations, which may introduce a touch of bias in the responses, but such a bias is easy to live with in the loose statistical framework of this discussion. Conversely, the term "advocate" has negative connotations to those scientists who treasure and live by the principle of total objectivity, so the absence of a plurality vote for this category may have resulted also from a soupçon of bias. I suspect, despite my own statistics, that there is more advocacy among scientists who are concerned with social issues than would be admitted publicly.

Those professionals who identify themselves as advocates on particular issues often have a clear rationale for their positions. Most of the major public issues of the day have strong scientific components, in that science-based technology is implicated, or that scientific data should form part of the background for decision-making. A listing of some principal issues of this kind include nuclear disarmament, pollution control, overpopulation, world food supplies and malnutrition, biological warfare, disease control, abortion, and global deforestation. Advocates feel that scientists and/or scientific information belong somewhere in the decision-making loop for every one of these issues, and should appear there. Advocates admit that responsibilities of scientists in public issues have been debated virgorously in all kinds of professional forums, and that while no real consensus exists, many see their roles as

- Ensuring public awareness of developing trends in data
- Clearly separating proven fact from interpretation or opinion
- Clearly indicating the level of certainty in any statement
- Using the most relevant statistical tests
- Recognizing that scientific advice is only one of several factors which may serve as bases for any policy decision
- Clearly summarizing the existing state of scientific knowledge on the topic of concern

Scientists who adopt advocacy positions on public issues must consider with great care how best to offer advice to public representatives (legislators, public health officers, environmental regulators, small town councils, and many others). The real danger comes in moving beyond, or being pushed beyond, the confines of available data, and making statements which cannot be supported adequately by those data. Professionals who can handle such encounters well, and still function within ethical boundaries, are important resources for all of us—scientists and nonscientists alike. To such advocates, the answer to the question "How much should scientists participate in commentary on public issues of the day?" is clearly this: "Participation in public forums in one's area of expertise should be part of the career expectation of every professional; we are part of a social system which maintains a significant niche for scientists and scientific activity—therefore, a contribution is expected and necessary."

Informed advocacy, then, can be an important adjunct to scientific accomplishments, enhancing the satisfactions of a career in research and teaching. If carefully circumscribed and executed, advocacy positions on public issues can enrich professional lives already characterized by technical successes.

THE EXPERT WITNESS

The scientific and legal universes intersect principally when scientists appear in courtooms or hearing rooms as expert witnesses. The stylized, almost ritualistic methodology of this intersection requires training, but is based on scientific credibility and good data—from which information relative to legal decisions can be extracted. Effective communication of technical information, in an adversary environment governed by rigid rules, is the objective. The extent to which the communication is successful depends on the skill and expertise of the scientist, alone in often-hostile surroundings.

In reviewing the case histories on which this book is based, it is remarkable how many excellent scientists have participated as professionals in legal proceedings of some kind— either courtroom cases or hearings. It is also remarkable that even though these were excellent scientists, most of them felt uneasy or disappointed about their initial performance in that specialized environment. A few who had repeat performances indicated that a rapidly accelerating learning curve existed, and that, after an initial cultural shock period, they were better prepared for subsequent appearances. They all felt the need for on-the-job training in courtroom procedures, and for better understanding of legal processes. They all offered tidbits of advice and rules to live by in the courtroom environment—often based on hard-earned lessons they had learned "on location." Many admitted to a feeling of exhilaration if performances were good, and some pointed out the satisfaction in participating actively in the application of scientific findings and conclusions to the decision-making process, especially in hearings related to public issues of the moment.

The role of the expert witness seems straightforward enough—responding to questions by disclosing the scientific facts and conclusions that are pertinent to the issue at hand. Unfortunately, like everything legal, the correct procedures and responses are not simple. Some admonitions and advice, gained mostly from notes in the case history files, are offered here in detail, partly at the insistence of those who had played early on as amateurs in a very hardball game:

- The role of the scientist as an expert witness is not a narrow one, since it may include several categories of participation:

  —In lawsuits and trials (including pretrial conferences and submission of depositions)—providing technical contributions to legal decisions in court cases
  —In administrative/regulatory hearings—quasi-judicial and leading to public policy decisions
  —As advisors to courts and other tribunals—providing objective analyses apart from the usual adversary role
  —As special hearing officers—who assemble information in advance of critical lawsuits at appellate levels

—In data gathering and presentation before international legal bodies, as, for example, in the recent United States/Canada boundary disputes, in which detailed environmental data formed the basis for an important legal position on both sides

- Expert witnesses provide facts, analyze available information, and make evaluations of the technical components of an issue—including expressions of professional opinions, if requested. Effective communication is a principal key, as it is in many human endeavors. The scientist has the information, but he or she must present it so that it will be accepted and used in evaluating any particular issue, and in reaching legal decisions.

- Expert witnesses should depend as far as possible on presentation of a carefully drafted *written* statement of facts and conclusions (as a "deposition" in pretrial hearings, or as a "statement" for an administrative/regulatory hearing). Where this is not feasible, every word of every oral statement must be carefully thought through, and no conclusions offered that cannot be fully supported by data, regardless of the provocation. Opinions and interpretations of data must be clearly labeled as such, and should be given only if requested to do so.

- Expert witnesses must be *competent*, with proper qualifications. This means that they must be speaking in their area of expertise, as outlined in a prepared statement of their training, degrees, experience, research, and publications.

- Expert witnesses must be *credible*; they should give the correct impression to the court or to hearing examiners that they are knowledgeable, informed, and articulate.

- Much of the credibility of a scientist depends on the ability to demonstrate that the facts he or she presents were acquired by standard methods or accepted investigational procedures. Core elements in establishing data quality include

—A demonstrable statistical basis for sampling and subsampling
—A satisfactory continuous control of samples taken for analysis
—Adequate replication of samples or experiments
—Use of standard analytic methods, especially if several data sets are used
—Intercalibration of methodology if several laboratories are working together
—Authentic and detailed notes on observations
—Acceptable confidence limits on conclusions reached through analyses of data

- Some of the usual courtroom rules—against expression of opinion and against conclusions drawn from indirect or secondary study—are waived for expert witnesses, but other rules of evidence (relevancy, for example) still apply. An overriding principle is that the rules by which scientists normally interact with each other do not apply in legal proceedings; failure to recognize this principle can lead to bruising encounters in a foreign environment.
- The weight of testimony of an expert witness is greater when he or she has firsthand information based on personal observation or examination, rather than on less direct study.
- Scientists should keep in mind that in the courtroom or hearing room they are the "performers," and they should seize opportunities to emphasize their legitimate strengths as expert witnesses, without becoming overly pedantic or arbitrary.

- Scientists and lawyers are similar in some fundamental ways (they are both pragmatists with high analytic abilities, and both depend on rules of evidence) but they are incompatible in other fundamental ways (scientists look for truth, whereas lawyers are merely trying to win cases, regardless of where the truth lies). Scientists would be well advised to treat all lawyers as intelligent humanoids from another planet, devoid of most human emotions and unpredictable in their actions or responses.

- An important point, often overlooked by experimental scientists, is that at which methods, findings, and concepts are acceptable to the community of science. There is an ill-defined boundary line which separates the experimental and unproven from the accepted stages in the evolution of scientific principles. Judges and other arbiters will depend on generally accepted procedures, theories, and statistical demonstrations in evaluating the validity of testimony given by expert witnesses.

- An earlier chapter on the Pathology of Science pointed out some of the consequences of adversary roles for scientists in legal proceedings—in which expert witnesses on either side may act to discredit colleagues. This is a real problem, unless strict ethical procedures are followed, and, even if they are, some loss accrues to science from such public disagreements.

- Expert witnesses who are also paid consultants move along a very thin ethical line. They are employed because someone thinks their testimony will support their employer's case, but they, as experts, can stretch conclusions only so far without losing credibility as scientists. Abuses are common, and major industries often develop a cadre of "pet scientists" who will invariably reach conclusions favorable to their employer. The extent to which scientific objectivity can be sacrificed to support particular positions (by biased interpretation of available data, by ignoring data unfavorable to a position or conclusion, or by denying the validity of opposing scientific opinions) should be matters of concern to all scientists, not only those who testify for hire.

So there sits the expert witness, alone, separated from supportive colleagues and from all the relevant literature, surrounded by members of an alien subculture with strange operating principles of partisanship, advocacy, trickery, and even deceit. To make a meaningful professional contribution in such a high-risk environment calls for the best in background and experience. For those who qualify, the rewards are substantial. The most important one has been mentioned already—the satisfaction of participating actively in the application of scientific information to decision-making, especially in hearings on public issues. Other less tangible rewards include the inner satisfaction derived from successful competition with very sharp legal minds, and a confirmation of credibility in forums far removed from "normal" science.

Representatives of the highly selected subpopulation of scientists who function effectively as expert witnesses were included in the larger sampling of professionals which served as source material for this book. They tended to be quick-witted and pragmatic, with excellent retention of facts and superb analytic abilities. Most of the long-term performers admitted to feelings of power and accomplishment after successful legal encounters.

THE POLITICAL LIAISON

Politics and politicians are important to science and scientists, since decisions about research funding and emphasis are influenced by political considerations. This state of affairs dictates that scientists (or their representatives) insert themselves in that part of the political process which concerns them, rather than waiting passively for good things to happen. An assertive role can be important in retaining a measure of control over government-funded research, and in reducing overt political intrusions into the conduct of science. The scientific–political relationship is always uneasy and unstable, but it is one with which all scientists must be concerned. A special contribution can be made by exceptional scientists who have achieved recognition in their fields. These are the people politicians will listen to; these are the people who should be doing and are doing the communicating.

A simple listing of ways in which scientists participate in activities that have political overtones may seem impressive to some scientists, and maybe disturbing to others who feel that science is above that sort of thing:

- Scientists and/or their organizations may participate in lobbying for or against particular regulations or legislation—an activity which stems from but transcends individual advocacy. (The current emphasis is on environmental protection and nuclear disarmament issues.)
- They may participate actively, or by membership only, in action groups for or against social issues of the moment (abortion, crime prevention, marine mammal protection, etc.).
- They may participate as members of advisory panels or committees created by government agencies or legislative subcommittees.
- They may appear as paid experts in individual or class-action legal suits against government agencies responsible for enforcement of regulations (as discussed in the previous section on "The Expert Witness").

- They may communicate directly with legislators or administration officials, as individuals or as representatives of groups, indicating approval or disapproval of actions, or requesting actions.
- They may develop personal relationships with legislative or agency staff members, and attempt to influence actions on programs or funding (a form of personal lobbying).

Probably obvious but still worth emphasizing is that scientists may interact directly with elected politicians and their staffs, or with appointed or career agency bureaucrats—in hearings, public information sessions, or advisory panels. Understanding the purpose and possible consequences of each kind of interaction is important to survival and happiness in a political climate, which is totally foreign to that of the everyday environment of science. Awareness of the nuances of political interplay is useful in contributing to a proceeding, to avoid harsh treatment from the regular participants. The most basic point to remember is that at the federal level, as well as at state and local levels, a three-way love–hate relationship exists among the administration, the legislature, and the career bureaucracy, with each vying for greater control.

The bureaucracy (in this instance the federal executive agencies) must be philosophically in accord with the administration in power. To ensure that this happens, political appointments are made as far down in each agency hierarchy as is feasible without drawing too much public reaction. For those agencies which have science components, such as the Environmental Protection Agency, the National Oceanic and Atmospheric Administration, and the Fish and Wildlife Service, this means that decisions about funding and the direction of science will be made by political appointees. They will endorse and follow policies of the administration in power (sometimes in opposition to their own personal views). Because of the transiency of administrations and their appointees, frequent reorganizations occur, policies change, and efficiency is hard to achieve. One truism exists, though, that the decision-making levels of any agency hierarchy will be occupied by political appointees, or by people responsible to an appointee for their positions. They direct the agency, which is otherwise staffed by career bureaucrats, who are nominally apolitical.

Scientists who engage in politically and emotionally charged issues of the moment that have social or economic consequences, such as genetic engineering or toxic waste disposal, must be prepared for sophisticated hardball encounters with politicians or their staffs. In these encounters, professionally based risk assessments are expected from the scientists, but value judgments from them are inappropriate. All the players—politicians and scientists alike—must (and usually do) recognize that data and conclusions drawn therefrom are the purview of science, but decisions based on value judgments (including those about the worth and utility of scientific information) are the province of the politician and not the scientist. This is true in part because scientific information is only one of many complex factors considered in most decisions and because power is in the hands of the politicians.

Among the many excellent credible scientists who elect to function at or near the scientific–political juncture, there is a prevailing feeling of urgency—of a need to translate scientific objectives and requirements into language comprehensible and acceptable to politicians. With research funding so much in government hands, effective relationships with bureaucrats as well as legislators can be critical to success in acquiring grants and contracts, or in funding new programs. The highly selected few who perform well at the border between science and politics constitute a special breed—to be nurtured carefully by the science community.

THE PUBLIC SCIENTIST

The continuity of support for scientific research today depends to a large extent on public funds, hence on public support. Public relations of scientists and scientific groups are terrible, except for small pockets of deliberately developed competence in communications within some research organizations. As a consequence, much of the interpretive work falls to the news media and to "public scientists" who often do a creditable job with the "gee-whiz" aspects of science, but usually fail to convey a rounded picture of scientific research and its value. Good science deserves a good public image; excellent scientists deserve public respect and recognition; but these worthwhile objectives are not easy to achieve.

The popularizers of science—Cousteau, Sagan, and their counterparts or imitators—clearly perform a useful function, purveying predigested concepts, minor propaganda spiels, and beautiful photography with excellent narratives. The factual material is usually well researched and presented interestingly. It all seems so glib and squeaky-clean, though, glossing over much of the grubby, repetitive, frustrating, sometimes maddening reality of scientific research. As such it tells only part of a story—the most superficially palatable part—so it denies to nonscientists a rounded concept of what science is all about. It might be fun, just once, to see a television documentary on a series of experiments that failed; all the animals got sick, some died, controls somehow got mixed with experimentals, the technician forgot to record critical data, and the results were in conflict with previous findings. It might be fun also to have a prominent scientist who literally stumbled into an important finding tell it like it really was, in public, and try to rationalize his approach with the so-called "scientific method." But—too bad—these events will not happen. Ratings might suffer, and the visuals would be difficult. Great things could be done with narration, though—if it could be done in typical understated British style—and an entire television series could be planned, with reasonable wit and substance, around real events in scientific existence. A marvelous idea—probably already proposed to and rejected by sensible producers.

Although there may be some wishful thinking on my part, it does seem that more excellent scientists are making serious efforts to communicate with people. Some go part way, presenting public lectures, participating in general-audience symposia, and even offering specialized adult education courses. Others are drawn into service club circuits, presenting interpretive sessions at Rotary, Kiwanis, Exchange, and other club meetings. Still others write for interested reader groups in magazines such as *Scientific American* and *Science 84*, or contribute to Sunday Supplement feature articles on scientific subjects.

All these attempts at communication by scientists seem to be paralleled by increased public interest, awareness, and sophistication. Well-researched news articles on scientific subjects seem more common in newspapers, and the range of scientific subjects covered in news stories is broad. Something comparable seems to be happening in television too, where science material is emerging from less-than-prime-time public television to a more visible place in commercial television. We may be seeing the consequences of greater concern about scientific advances—a reasonable progression since science is involved to a significant degree in many events that are at the center of public interest. Nuclear war, genetic engineering, overpopulation, famine, environmental degradation, abortion, disease control, biological warfare, medical advances—all are matters with a principal scientific component.

CONCLUSIONS

Here than—in the Advocate, the Expert Witness, the Political Liaison, and the Public Scientist—are just a few examples of the supralaboratory roles which may be assumed by excellent scientists, usually those well along in successful technical careers. Other examples might be those who act as science advisors, as consultants, or as commentators on science affairs. Discussions with those who participate in these externally oriented activities disclosed satisfactions and pleasures which may equal or even transcend those found in research and teaching. Testifying in an area of expertise before a Congressional committee; appearing as an expert witness in an important court case; developing critical data contributing to decisions on environmental issues; transforming complex technical information into summaries useful to public understanding—these are some of the sources of joy for those who elect to take those additional steps, to expand their horizions beyond the secure but restricted environment of research and teaching.

Scientists can provide rational voices in a period of history characterized by strongly held opinions on almost every social issue. The joys in the role are many: presenting carefully collected and analyzed data as one basis for decisions on public issues; occasionally taking the lead in public action if the facts clearly indicate the necessity; and even adopting the self-perception that scientists constitute a thin line of realism and fact—a class of public defenders difficult to subvert—drawn up against the manipulative forces which operate for selfish purposes.

Of course, not all elect this kind of outreach function; many prefer to invest totally in affairs within the community of science and are happy with career destinations described early in Chapter 6—specifically the Research Scientist and the Scientist-Educator. These are the ones who write textbooks, edit journals, organize symposia and workshops, and act as motive forces in scientific societies. But whatever the chosen route or combination of routes, excellent professionals contribute to the resolution of important science-related sociopolitical issues of the times, principally by supplying data and interpretations of data, but also on occasion by active participation in the decision-making process.

EPILOGUE

This book on "The Joy of Science" is based on information gleaned from prolonged discussions with a large sample of excellent scientists—whose anonymity has been preserved, but whose traits have been variously categorized, analyzed, and maybe even romanticized. An underlying thesis throughout the narrative has been that excellence in science is recognizable, definable, and attainable—and that the pleasures of a scientific career are commensurate with the successes. While such a thesis may appear to be a little simplistic, it is one which heretofore has had remarkable underemphasis in career descriptions of a large population of productive professionals.

The extent to which this document conveys some of the joys of "doing good science"—as distilled from many case histories of successful practitioners—depends of course on perceptions of each reader. An attempt has been made to inject a degree of realism into the discussion through consideration in Part Three of mediocrity and failure, but a clear and intentional bias toward the positive exists in most of the text. Excellent scientists enjoy what they do; most feel that theirs is the best of all occupations. This book tries hard to describe some of their joy. Even in a scientific era when the employment scene is less than perfect, those who persist and produce will prosper.

INDEX